THE MANUAL OF **PLANT GRAFTING**

THE MANUAL OF

PLANT GRAFTING

Practical
techniques
for ornamentals,
vegetables,
and fruit

PETER T. MacDONALD

TIMBER PRESS | Portland · London

For Elizabeth, Eilidh, and Alastair

Illustrations by Kate Francis.
Photo credits appear on page 214.

Published in 2014 by Timber Press, Inc.

The Haseltine Building
133 S.W. Second Avenue, Suite 450
Portland, Oregon 97204-3527
timberpress.com

6a Lonsdale Road
London NW6 6RD
timberpress.co.uk

Printed in China
Cover and text design by Susan Applegate

Library of Congress Cataloging-in-Publication Data

MacDonald, Peter T.
The manual of plant grafting: practical techniques for ornamentals, vegetables, and fruit/Peter T. MacDonald.—First edition.
     pages cm
Includes index.
ISBN 978-1-60469-463-5
1. Grafting. I. Title.
SB123.65.M23 2014
634′.04—dc23   2014009482

A catalog record for this book is also available from the British Library.

CONTENTS

Acknowledgements

THIS BOOK COULD NOT HAVE BEEN WRITTEN without the generosity of time and knowledge from many growers. I would like to thank the following and their staff: Nick Dunn of Trees for Life (Frank P. Matthews Nursery), John Richards of John Richards Nursery, Roger Ward of Golden Grove Nursery, Mike Norris and Andy Mount of New Place Nursery, Stein Berg of Yorkshire Plants, Chris Lane of Witch Hazel Nursery, Richard Merritt of Plant Raisers, David Hide of Fargro, Adam Train of Train's Nursery, and Carlos Verhelst. I hope that this book does justice to their skills and knowledge.

I am grateful to my colleagues at Scotland's Rural College who commented on aspects of the text, checked plant names, and helped with typing: Ian Cornforth, Leslie Gechie, George Gilchrist, Angela Lloyd, Mark McQuilken, Jeanne Morton, and Ian Ratchford. I would also like to thank Phil Lusby at the Royal Botanic Gardens, Edinburgh, for checking plant names.

In addition, I would like to thank Rev. Fraser Aitken of Ayr St. Columba Church for his comments on religious views on grafting, Stephanie Dalley of the University of Oxford for clarifying issues relating to cuneiform references to vine propagation, and Heinrich Loesing for photographs of using budding to test for disease resistance. I am very grateful to Brian McDonough for much of the information and photographs about cactus grafting. I am also grateful to Nick Dunn, Steen Berg (Yorkshire Plants), Jill Vaughan (www. organicplants.co.uk), Juliet Day, ISO Group, and Alastair Swan for use of their photographs.

I would like to acknowledge the members of the wonderful International Plant Propagators' Society for their commitment to "Seek and Share" and the wealth of insight within the *Combined Proceedings* from the Society's conferences.

I would also like to thank Anna Mumford and Linda Willms of Timber Press for their help and expertise in the production of this book.

Lastly, but certainly not least, I would like to thank my family for their encouragement and help with this book. In particular, to Elizabeth for correcting my English and being so supportive, to Eilidh for accompanying me on a plant hunting trip to deepest Ayrshire, and to Alastair for his encouragement.

INTRODUCTION

GRAFTING, THE ACT OF UNITING PART OF one plant with another so that they become a single plant, has been used as a method of propagation for several thousand years. The original purpose of grafting was to propagate plants vegetatively, that is, to produce plants that are genetically identical to the parent plant. Over the years, however, other benefits have been developed from joining the roots of one plant and the top growth of another. Grafting is particularly important for pest, disease, and vigour control, but can be used for other reasons.

The development of grafting has involved many people. Some are famous such as Virgil, Alexander the Great, and Charles Darwin. Others such as Philipp Franz von Siebold and Frederick Burbidge are well known in horticulture, although their story may not be familiar. Still others such as Jules Planchon and Charles Riley are little known but deserving of more recognition. In this book, the history and uses are described, showing how nearly all the grafting methods used today had been developed by the end of the nineteenth century. Not until the twentieth century did the science of grafting start to be properly understood—although there is still much to learn. Understanding the scientific principles of grafting has helped to improve and develop the grafting techniques commonly used today. These principles are outlined in this book to provide the underpinning knowledge required to succeed in grafting a plant.

One of the most important requirements for a successful, good-quality grafted plant is the quality of the initial rootstock and scion. The main methods of producing rootstocks are also described—by seed, layering, and cuttings. In practice, however, most people who graft plants purchase their rootstocks from specialist producers. It is important that they are able to obtain plants of the correct specifications, so the requirements for different types of rootstock are provided. The provision of the correct scion material, principally by pruning, is also discussed.

On visiting a number of growers, it was obvious to me that there are many ways of grafting most plants. These may be large differences, such as between budding (attaching a single bud to the rootstock in the field) and grafting (attaching a bud-stick of three or four buds to the rootstock on the bench). Other differences are less obvious, such as waxing or not waxing a side veneer graft.

There is no single, correct way to graft a plant. There are, however, different ways of successfully grafting. These are not necessarily preferred or better—just different. Therefore, it is not possible to provide one technique for the grafting of each species, there are simply too many options available.

In the chapter on bench grafting, the grafting approaches have been divided into cold and hot callus grafting. The choice of one or the other depends on the time of year and whether artificial heat is applied to the graft. Although growers use many different actual grafts, all of the grafts fall into two types: apical grafts (where the top of the rootstock is cut back prior to grafting) and side grafts (where the top of the rootstock continues to grow above the graft union for a time). This book includes examples and details for managing the rootstocks and grafts after the union.

After the quality of the original plant material and the timing and aftercare of the graft, the third key element to success is the craft skill required to make the graft cuts. How to prepare an apical graft and a side graft is described. If the described techniques are mastered, then it is possible to adapt the use of the knife for any other specific graft required.

Budding in the field is the other main method of grafting used by growers. The two methods of budding described in this book are T and chip. The reasons for using each and the process of successfully carrying out budding are explained.

Most of the book concentrates on woody plants, but the grafting of tomatoes and other vegetable salad crops is becoming increasingly widely used. Although the possibility of grafting these types of plants has been known since ancient times, it has been the Japanese in particular who have led to its popularity in recent years. The advantages of grafted plants are discussed and the methods used for a range of species outlined. Cacti can also be grafted using a technique different from that used for woody and vegetable plants. The grafting of cacti is, therefore, given a separate chapter.

What is the future of grafting? It is always difficult to predict the future, but some of the possible developments and uses are discussed. Finally, at the end of the book are three charts of woody plants, both ornamental and fruit, that may be grafted. If you are looking to graft a particular plant, the charts suggest options for the grafting method to be used and suitable rootstocks.

One of the main aims of this book is to discuss in detail the principle techniques being used by growers. I have been fortunate to go on study tours to the United States, the Netherlands, Germany, and Australia. These have enabled me to see something of their grafting methods. For the majority of the information on practical grafting, however, I have had the assistance of many propagators working on nurseries in England that specialize in grafting. Their location in England should be borne in mind when considering the timings and specific details of the grafting techniques, especially the aftercare. Minimum temperatures in England are -7 to $-12°C$, approximately equivalent to USDA Hardiness Zone 8. Although the nurseries lie between latitudes 49 and $52°$ North, the British Isles have a maritime climate and therefore do not have extremes of temperatures. The average minimum winter temperature is $1.6°C$ and an average maximum summer temperature of $20.9°C$. In midsummer, day length is 16.5 hours with an average of 6.5 hours sunshine a day. In midwinter, there are only 8 hours of daylight with an average of 1.5 hours of sunshine a day.

The other principle source of information for the practical application of grafting has been the journal, *The Combined Proceedings of the International Plant Propagators' Society*. Any professional horticulturist involved in producing plants should be a member of this society. The journal goes back over 50 years and holds a wealth of knowledge on all aspects of propagating and growing plants. More importantly, with the motto "Seek and Share," its members freely exchange knowledge, making it a very friendly and supportive society with which to get involved.

If you are new to grafting, I hope this book will give you the confidence to have a go. If you already graft, I hope you will find a few pointers to help you improve your success rate or quality of final plant. If you just have an interest in gardening, I hope you will be inspired to find out more about some of the characters who have contributed to the development of grafting over the years.

1 History of Grafting

BETWEEN THREE AND FIVE THOUSAND years ago, a farmer took a shoot (or scion) from a plant and attached it to another plant (or rootstock) growing nearby in such a way that they formed a union and the shoot began to grow. The first graft had been successfully carried out.

To achieve this, however, the two plants had to be related closely enough to be compatible and form at least a temporary union. A cut would need to have been made on both plants and put together so that vascular cambium cells were close enough to form a connection across the callus bridge. The callus bridge would only form if the two plants were held together and prevented from drying out. The vascular cambium would only form if the tie were tight enough to apply some pressure to the cuts. Finally, the entire process would only be successful if done at the right time of year when cells were actively dividing in the rootstock and the scion buds were dormant.

How many times might this have been tried before a successful union was achieved? How often would someone persevere in trying to achieve a union if the first attempt was unsuccessful? Propagating a new plant by grafting does not seem an obvious thing to do, and it is certainly not a straightforward propagation method. In contrast, it is quite easy to imagine how the cultivation of crops from seed might have developed. It would only require hunter-gatherers to drop some seeds around a campsite as they returned from a foraging trip. In time, it would be evident that collecting seeds from the new plants would be far easier than going out foraging. Offsets from date palms could be dug up and replanted from the parent plant as a simple method of vegetative propagation. Stems of trees might have been put in the ground to keep in animals. Some of them could have rooted and led to people deliberately taking simple cuttings.

Grafting, however, is a method of propagation that requires skill to make the cuts correctly, to tie the parts together with the precise match of cuts, to right amount of pressure, and to ensure the cuts are properly sealed. So how and why did these hunter-gatherers develop the skills to do it?

Grafting in nature

Inosculation, the act of uniting or being united so as to be continuous, has been used to describe two plants or plant stems growing together and their cambiums fusing to form a union. This phenomenon occurs relatively often in nature. On the banks of the river Ayr, for example, next to the campus of Scotland's Rural College, a branch of sycamore (*Acer pseudoplatanus*) has become attached to the trunk of a larch (*Larix decidua*). Although a union has formed, it cannot be described as a natural

graft since, if the roots of either plant were severed, the top growth would soon die.

Inosculation of roots seems to be quite common between woodland species, and there are reports of interspecific unions between different species of oak (*Quercus*), maple (*Acer*), and pine (*Pinus*). The formation of these interspecific unions may explain some of the graft combinations written about by early authors, but they could not have actually produced permanent grafts.

Natural grafting is a form of inosculation, but occurs between plants that are capable of forming a union that will support one or the other if a root system is severed. Some natural grafts are quite common, such as between sycamore (*Acer pseudoplatanus*) and ivy (*Hedera helix*) or within beech hedges (*Fagus sylvaticus*) where the dense tangle of stems can create the correct conditions for grafting.

Natural grafts form when branches grow against one another. Although the plant stems are not cut, slight movement as the stems grow together will cause the bark to wear and expose the cambium, allowing the graft process to occur. When plants are woven together, natural graft unions also occur. This can be seen in the weeping fig (*Ficus benjamina*) popular in shopping centres. These often have three stems pleated together that gradually form a graft union as the stems thicken and grow tightly together. Woven structures made from willow (*Salix caprea*) will also form grafts as they grow together. Other species where natural graft unions have been observed include *Acer*, *Magnolia*, *Prunus*, and *Quercus*. Although grafting is often said to be a technique that would have developed later than most other vegetative methods, it is not that different from the other techniques in that it would have been developed through observation.

Inosculation: A branch of sycamore (*Acer pseudoplatanus*) has formed a union with the trunk of a larch (*Larix decidua*) on the banks of the river Ayr.

Pleated weeping figs (*Ficus benjamina*) are available in garden centres. The stems fuse together to form a natural graft.

Evidence for earliest grafting

To answer the question about why early humans used grafting, we must consider their changing behaviour. Around thirteen thousand years ago, the evolution of humans from hunter-gatherers to settled communities was beginning. Human behaviour began to change due to the climate becoming unpredictable, to decreasing big game that had been the main food source, and to increasing population pressure.

People started to change their diets to smaller game animals and plant foods that required more preparation like grinding, leaching, and soaking. Eventually, people transported some wild plants (such as wild cereals) from their natural habitats to more productive habitats and so began

Natural grafting is common in ivy (*Hedera helix*). The stems are all interconnected and will continue to grow even if several stems at the base are cut through.

intentional cultivation and the development of settled communities.

It is now known that, initially, human health did not improve in settled communities, and nomadic hunter-gatherers were likely to have healthier lives than those dependent on farming. There were, however, other strong incentives for groups or tribes to stay in fixed locations: the safety and protection that village communities could provide; the opportunity to develop technologies, and make and acquire goods; and the chance to maximize resources like grain stores to help large groups of people through lean times.

The benefits of settled communities were, therefore, the reason for the development of agriculture and horticulture, rather than the other way round. But settled communities could not occur anywhere, as successful communities were only sustainable where suitable plants and animals for domestication existed. It has been estimated that there are nine centres of plant domestication: the Fertile Crescent, China, Mesoamerica, Andes/ Amazonia, eastern United States, Sahel, tropical West Africa, Ethiopia, and New Guinea. Written and archeological evidence shows that grafting originated in either the Fertile Crescent or China.

The Fertile Crescent

Domestication first occurred in the Fertile Crescent, which stretched in an arc from the Nile to the Tigris and Euphrates rivers. A large number of our domesticated plant and animal species originate from this area: wheat, barley, peas, dates, olive, figs, grape, pomegranate, sheep, goats, cows, and pigs.

Once communities became established, it became necessary to develop cultivated crops to feed them. This quickly led to what has been called the domestication syndrome—the rapid divergence of wild and domesticated plants as they were selected for characteristics such as

higher germination rates, more even germination, increased size of seed, and ripe seeds that stay on the plant (rather than breaking off and falling to the ground).

Annual crops including wheat, barley, and peas have the advantage of limited heterozygosity of genes. This lack of genetic variation allows these plants to inbreed, which means that they will come reasonably true from seed and have consistent characteristics. The newly settled communities would also encourage the investment in long-term perennial crops. The cultivation of dates, olive, grape, figs, and pomegranate occurred in the Fertile Crescent from about 4000 BCE. Woody perennial plants are less consistent from seed, especially if specific fruit characteristics are required. Most fruit crops are very heterozygous, and so any plants raised from seed will show much variation. To select varieties of fruit rather than just collect plants from wild sources, communities needed to develop vegetative propagation techniques such as division, layering, and simple cuttings. These would soon have been developed and have been used to propagate these fruits.

The earliest date claimed for grafting in the Fertile Crescent comes from cuneiform tablets written in Mesopotamia about 1800 BCE. It has been claimed that the tablets describe bud-sticks used for grafting. In fact, they only describe grapevine shoots for the vegetative propagation of vines and probably refer to propagation by cuttings.

Texts in the Bible are also said to provide evidence for grafting. Leviticus, for example, written around 1400 BCE, gives instructions to the priesthood and in 19:19 states:

> Ye shall keep my statutes. Thou shalt not let thy cattle gender with a diverse kind: thou shalt not sow thy field with mingled seed: neither shall a garment mingled of linen and woolen come upon thee. (King James Bible)

Scholars have argued that the mingled seed in this text includes grafting, which would have been seen as mixing plants together. Support for this position is said to be in the Jewish Mishna Torah, written in 300 CE, which provides further instructions on purity and states: "It is unlawful to graft tree on tree, vegetable on vegetable, tree on vegetable, or vegetable on tree" if they belong to different species. This later instruction was written long after grafting was known to be practised and does not necessarily support the argument that Leviticus did refer to grafting.

Moses Maimonides (1135–1204), a Jewish philosopher and Torah scholar who is still important for his comments on Talmudic law, also commented on *Nabatean Agriculture*, a book written between the third and ninth centuries, which deals not only with agriculture but also with magic and sorcery. In his book *Guide for the Perplexed*, Maimonides wrote, "I have already shown and explained to you how the Law opposes all kinds of witchcraft." He goes on to write about grafting:

> Another belief that was very common in those days and that survived the Sabeans is this: when a tree is grafted into another in the time of a certain conjunction of sun and moon, and is fumigated with certain substances whilst a formula is uttered, that tree will produce a thing that will be found exceedingly useful.

Since the Sabeans were a people believed to predate Adam, this suggests an ancient tradition of grafting. More extraordinary are some of the grafting techniques attributed to the Sabeans:

> More general than anything mentioned by the heathen writers was the ceremony of grafting an olive branch upon a citron tree, as described in the beginning of *Nabatean Agriculture*. . . . [T]he branch that is to be grafted must be in the hand of a beautiful damsel, whilst a male person has disgraceful and unnatural sexual intercourse with her: during that intercourse the woman grafts the branch into the tree.

Maimonides's conclusion gives some explanation why mingled seed and grafting are linked together:

> The Law, therefore, prohibits us to mix different species together, i.e., to graft one tree into another, because we must keep away from the opinions of idolaters and the abominations of their unnatural sexual intercourse. In order to guard against the grafting of trees, we are forbidden to sow any two kinds of seed together or near each other.

Biblical texts also refer to the reversion of cultivated plants to their wild form. Jeremiah 2: 21 states:

> Yet I had planted thee a noble vine, wholly a right seed: how then art thou turned into the degenerate plant of a strange vine unto me? (King James Bible)

Once again, it is claimed that this text refers to a grafted vine reverting to a wild form due to suckering from the rootstock. In fact, the text seems to refer to the use of seed collected from cultivated plants that show unwanted variation, rather than the use of plants propagated by cuttings or other vegetative means that would retain the cultivated characteristics.

Until well into the Common Era, biblical texts do not refer specifically to grafting, and any reference to growing plants can be more easily explained by propagation methods other than grafting. Although grafting is the normal method of propagation today for fruits like grapes, it is a relatively new development and would not have been a requirement of vegetative propagation in ancient times.

China

Archeological evidence suggests that agriculture developed later in China than in the Fertile Crescent. Since there are strong benefits in the development of settled communities, it may be that agriculture arose independently in China. Alternatively, knowledge of agriculture and the domestication of crops could have been spread through Mesopotamia, India, and on to China. There is evidence of an ancient Neolithic trade in flint supporting the hypothesis of the movement of trade and knowledge leading to a spread of domestication. Whatever the truth, Central Asia around northwestern China was where pome and stone fruits originated, while in Southeast Asia, specifically southeastern China, peach, citrus, banana, and mango (among others) originated. These became some of the most important fruits in cultivation.

Certainly, grafting is required to maintain clones of many of the fruits originating from China and there are claims that it was being used by 5000 BCE. According to a reference to *The Precious Book of Enrichment*, written around this time, "Feng Li [a Chinese diplomat] gives up his position when he becomes consumed by grafting peaches, almonds, persimmons, pears and apples as a commercial venture." While this book is often quoted, the quotations all seem to come from a single source for both the book and Feng Li. It has not been possible to find the book, even in the British or Congressional Libraries.

In the sixth century CE, Chinese author Jia Sixie wrote *Qimin Yaoshu* (Essential Skills for the Common People), a 10-volume book about agriculture, forestry, and fishing. In this, he referred to an earlier lost book that included information on grafting, suggesting that this mention showed that grafting was first used when Da Yu was emperor

(2205–2197 BCE). Again, however, the earlier text is missing and so this early date cannot be substantiated.

The domestication of apples occurred from 5000 to 3000 BCE in the Central Asia area of modern-day Kyrgyzstan, Kazakhstan, and Xinjiang in northwestern China, and spread from this area along the silk trade routes. Gene analysis has identified *Malus sieversii*, originating in this part of Central Asia and still found there to this day, as the ancestor of the present-day domesticated apple (*Malus ×domestica*). Clones of this fruit tree would have to have been fixed by propagating through grafting, as it is not easily propagated by other vegetative means. However, the domestic apple can be grown from seed and, although fruit quality would vary, it would still be an improvement over the wild forms. In fact, even in the nineteenth century, the famous American Johnny Appleseed (Jonathan Chapman), a nurseryman who introduced the apple to large parts of the eastern United States, grew his trees from seed as he believed this gave better fruit for cider and animal feed than clonal plants from grafting. There is, therefore, a strong argument to be made that grafting originated in China, but no absolute proof.

Greece, Rome, and Islam's Golden Age

The earliest definite written account of grafting known today was made by one of the followers of Hippocrates in Greece in 424 BCE. Pseudo-Hippocrates stated that "some trees however, grow from grafts implanted into other trees; they live independently on these and the fruit which they bear is different from that of the tree on which they are grafted." The writer did not imply that grafting was something new; it seems to have been a well-established technique before this time. By 300 BCE, Greek writers described the use of rootstocks to control the vigour of trees, including dwarf apple trees.

Several different grafting techniques were used at the time, and in the last two centuries BCE, writers in Rome described budding, cleft, and approach grafting, suggesting that grafting techniques for different species had become quite sophisticated. Approach grafting involves bringing together two plants growing on their own roots, grafting them together, and then severing the roots of the plant that is to be the scion. While intuitively it may seem that this would have been the earliest method of grafting, being adapted from observing what happened in nature, one author, Marcus Terrentius Varro, suggested that it was a recently developed technique at this time. It may be that he was describing a particular method of approach grafting, rather than approach grafting as a whole, being new. While written references to grafting occur from 500 BCE, the oldest known image of grafting is in a Roman mosaic from 300 CE in St. Romain-en-Gal in Vienne, France, and shows detached scion grafting.

From the fifth century, following the collapse of the Roman Empire, Europe entered a period of intellectual and economic regression. The Islamic Golden Age from the eighth century was when the Arab world became the intellectual centre for science, philosophy, medicine, and education. One important aspect of this period was the development of libraries like the great library of Alexandria, which preserved many of the earlier Greek and Roman texts. These provided the sources that led to the advancement of science and education until the Mongol invasions of the thirteenth century. Of particular social importance during this time was the interest in ornamental gardens, with rulers in Basra, Cairo, Damascus, Seville, Cordoba, and Valencia competing for the best gardens. These would have ensured that horticulture knowledge and skills were retained and developed.

Europe and the United States

Apart from the immediate period following the fall of the Roman Empire, Europe did progress, the first universities were established, and agricultural surpluses were produced due to a period of excellent weather. The monasteries were of particular importance in retaining horticulture skills and knowledge. Albertus Magnus (1200–1280), a German Dominican friar and Catholic saint, is particularly noteworthy. He studied the writings of Aristotle and wrote many books, in particular *Physica*. He was one of the first to recognize monocots and dicots, vascular and nonvascular plants. He also wrote about grafting, believing it was a means to improve cultivars and not just a means of propagation.

Horticultural knowledge in Europe would make significant progress again, following the Renaissance and the availability of gardening books with the arrival of the printing press. Commercial production started to become important in the sixteenth century. The Low Countries, in particular, developed their horticultural industry early and exported significant amounts of seed to England by the mid-sixteenth century. The first commercial nurseryman in the London area was John Banbury, a basket maker who probably started the nursery to grow his own willows. In his will of 1560, he instructed his son Henry "to plante and grafte for the behalf of his mother and he to have the third part of his labour." It would seem that Henry carried out his father's wishes as John Gerard commented in his herbal that Henry was "an excellent graffer and painful planter, Mr Henry Banbury of Touthill Street neere unto Westminster." At this time in history *painful* meant industrious.

In 1683, *The Scots Gard'ner*, the first Scottish gardening book, was written by John Reid. In it, wedge graft, saddle graft, top working, approach grafting, root grafting, and T budding are described. "There are many other wayes, but these nam'd are the most material" wrote Reid.

The history of plant collecting began centuries before Europeans engaged in the activity. Queen Hatshepsut of Egypt sent botanists to Somalia to collect incense trees about 1495 BCE. Once European explorers started to open up trading routes with Asia and the Americas, plant collecting became a major pastime. By the Victorian Age, large amounts of money were being invested to send plant collectors such as Robert Fortune, David Douglas, and Thomas and William Lobb on collecting expeditions. Their efforts yielded numerous new plants, many of which required to be propagated by grafting. The introduction of cultivated roses in the eighteenth century, for example, gave the first repeat-flowering roses in

Europe as well as a wider range of colours and cold resistance.

Many hybrid tea rose cultivars were soon being grown but growth was poor due to weak, spindly roots. This problem was overcome by the use of *Rosa multiflora*, a vigorous plant with nondescript flowers, as a rootstock. By Victorian times, rose standards (bush roses budded on the top of a tall rootstock such as *R. rugosa*) were a particular favourite in municipal parks.

By the end of the nineteenth century, almost all the grafting techniques used today had evolved. The only significant new technique of the twentieth century has been *in vitro* micrografting for the production of disease-free plants. The large number of techniques available to the grafter can be seen in *The Grafter's Handbook* by R. J.

Visitors enjoy roses at Keisei Nursery Rose Garden near Tokyo, Japan.

Garner. In the twentieth century, commercial plant propagation has tended to reduce these to only seven or eight commonly used techniques. This reduction is due in part to technical developments and in part to the need of propagators to use the quickest, most straightforward methods possible. Chip budding, for example, is not a new method of grafting but only became a common technique for budding trees in the latter part of the twentieth century with the development of polythene tape.

In the late 1970s, Harry Lagerstedt, instructor of horticulture at Oregon State University, began to develop the callus hot-pipe to improve the success of winter bench grafting of deciduous trees. Verl Holden of Holden Wholesale Growers in Oregon took up this idea and made it into a

Rose Garden at the Royal Horticultural Society's garden at Wisley showing the use of standard roses to give height to a rose bed. Produced by budding shrub roses onto a tall rootstock stem.

commercial technique. Through the International Plant Propagators' Society, hot-pipe technology has spread quickly and has been adopted widely by specialist propagators.

Around the same time that these developments were occurring in grafting, progress was made in rooting cuttings. As simply a propagation technique, grafting has therefore declined as developments in mist, polythene, fog, and the heated propagation bin have made it possible to root a greater range of plants from cuttings. For other reasons, however, the use of grafting has not declined, as it is increasingly important for other applications like the production of unusual growth forms or the benefits gained from particular rootstocks.

Benefits of grafting

While grafting was used primarily as a method of vegetative propagation to maintain the characteristics of particular clones, other benefits were soon recognized. Jia Sixie, the sixth-century Chinese author, describes an early method of grafting bottle gourd that may have been done to give resistance to soil diseases or to increase vigour. In this case, ten seeds were planted in a hole and the emerging shoots tied together so they fused (approach grafting). The plant was pruned to a single shoot attached to the ten roots. The benefits of this technique may have been to delay root death due to fungal diseases, increase gourd size, or to extend the length of the season due to the amount of root growth occurring from the ten plants initially sown.

Jia also described how grafting could be used to manipulate the growth of trees. Referring to *Pyrus pyrifolia*, the pear species native to China, Japan, and Korea that produces large, round fruit, he explained that for grafted trees to be grown in a courtyard, the scion should be inserted upwards to "make no trouble with the building," while for

Ten different culinary pears (*Pyrus communis*) grafted onto a single rootstock at the Aalsmeer Horticulture Museum, Netherlands.

Multigeneric grafting of conifers. Two different conifer scions are attached to a single rootstock to produce an unusual plant form.

grafted pears to be grown in an orchard, the scion should be inserted sideways to make picking the fruit easier.

Alexander the Great (356–323 BCE) conquered much of Asia as far as northwestern India and is believed to have introduced dwarf fruit trees, mentioned by Theophrastus. Although these trees may not have been used for grafting at that time, it is thought that they may be part of the parentage of the modern dwarfing apple rootstock Malling 8 (M8). The dwarfing rootstock 'Paradise' was known in England in the fifteenth century, but probably dates back earlier to France. There was actually more than one clone of 'Paradise' giving differing vigour control which could be propagated from cuttings. In the early twentieth century, East Malling Research Station collected a wide range of apple rootstock material and began a breeding programme that has produced several rootstocks that give vigour and pest and disease control. The use of grafting to produce unusual

forms of plants like "family" fruit trees (where several varieties are grown on one rootstock) was probably made quite early on as people experimented with grafting.

Grafting to control pests and diseases became important in the nineteenth century when phylloxera (*Daktulosphaira vitifoliae*), an insect originating in North America, threatened the French wine industry. Citrus grafting also began around this time to control gummosis, a root rot caused by various species of *Phytophthora*.

Frederick William Burbidge was a plant collector who worked for the famous Veitch Nurseries of Exeter, the largest group of family-run nurseries in Europe in the nineteenth century, introducing around thirteen hundred plants into cultivation. Burbidge travelled extensively in Southeast Asia and introduced the pitcher plant (*Nepenthes rajah*) into cultivation from Borneo. The genus *Burbidgea*, a member of the ginger family, is named in his honour. Writing in

Apple trees at Brogdale, home to the United Kingdom's National Fruit Collection, showing the use of dwarfing rootstocks to control the vigour of tree growth.

Cultivated Plants: Their Propagation and Improvement in 1877, Burbidge described the use of an interstock when grafting:

"Double grafting" is a comparatively modern practice, principally resorted to in the case of delicate-rooted varieties of the Pear, of which such as Seckel, Van Mons, Beurre Flou, Huyshe's new varieties, and others, may be cited as examples. Some Pears do not bear well on their own roots, neither will they flourish on the Quince stock; yet when some strong-constitutioned Pear is worked on the Quince, and used as an intermediate or go-between stock for a delicate variety, or one which does not coalesce perfectly with the Quince stock, excellent results are obtained.

Although it was not understood at this time, the technique described was to overcome "localized incompatibility" that occurs with some culinary pear varieties grafted onto quince rootstocks.

In the twentieth century, the grafting of vegetables began in Japan to overcome soil-borne diseases. Grafting for more than just propagation has therefore expanded and produced a large number of applications that will be discussed further in chapter 2.

Developing the science of grafting

While the craft of grafting techniques developed to give a successful method of propagation, the understanding of the science behind the technique developed more slowly and is still debated.

The Japanese apple pear (*Pyrus pyrifolia*) just prior to harvest. Many fruit trees naturally shed fruitlets in June. Once this has occurred, the remaining fruit are covered in paper bags to prevent any damage to the skins of the pears and to keep the fruit perfect.

In particular, three aspects of grafting have been debated over several centuries. The first of these is incompatibility and an understanding of the limitations of graft scion combinations. Many practical horticulturists must have been disappointed with results if they followed the instructions of some authors, right up to the seventeenth century. Secondly, there was the attempt to understand grafting in relation to the theories of plant growth and in particular the paradox of "Specific Fluid." Finally, there is the ongoing debate about whether graft hybridization occurs or not and the importance of this debate as far as the acceptance of grafted plants in some communities is concerned.

The principle of incompatibility

Writing between 371 and 287 BCE, Greek author Theophrastus showed that many of the requirements for successful grafting were understood by this time. He wrote that desiccation was avoided by "cutting stock and scion precisely so they fit together tightly and the core is not exposed to drying." The joint was then to be held together by layers of lime bark, plastered mud, and hair. In some cases, a pot of water was to be set over the graft and the water allowed to drip on it.

Theophrastus showed some understanding that not all plants would graft together when he instructed that the stock and scion were to have "like bark." He also stated that the time of bud break was important. His writings describe the need for cambial contact, pressure, moisture, compatibility, and correct timing, and give evidence that such practices were being carried out.

Selecting the correct rootstock and scion combination is important as not all combinations are compatible and therefore do not give a permanent union. Theophrastus showed some understanding of this when writing about "like bark," but many other authors recommended scion-rootstock combinations that showed a lack of first-hand knowledge of incompatibility issues.

Virgil (70–19 BCE), author of the *Aeneid*, also wrote texts on agriculture that included instructions on grafting. Among the graft combinations he discussed are arbutus and beech with walnut, mountain-ash with pear, and oak with elm. These pairings show that he knew of grafting but did not have first-hand experience with these incompatible combinations, unless inosculation was used.

Similarly, Pliny the Elder (23–79 CE) wrote correctly about grafting grape vines, but went on to give a fanciful description of natural grafting by seed. In this, he described how a "famished bird" will swallow seed whole and softened in the crop to be deposited on a branch of a tree with the "fecundating juices of the dung." This process allegedly gave rise to cherry growing upon willow, laurel growing upon cherry, and "fruits of various tints and hues all springing from the same tree at once."

By 400 CE, Rutilius Taurus Aemilianus Palladius had written *Opus Agriculturae*, a fourteen-volume book that included a section on grafting. He described a range of suitable scion and rootstock combinations and, while many were correct, like olive on wild olive and pear on quince, many combinations were in fact incompatible, like plum with chestnut, suggesting yet again an author's lack of first-hand experience.

From the fifteenth century, the Renaissance began in Italy and spread throughout Europe. It is a period famous for such talents as Michelangelo and Leonardo de Vinci, and saw a resurgence of learning based on classical sources and the beginnings of scientific discipline in trying to understand the natural world through observation. The increase in literacy and the invention of the printing press at this time helped spread cultural and scientific knowledge around Europe, and many gardening books started to be written. It would be expected that the new approach to science by observation would prevent some of the errors in rootstock-scion combinations that appeared in Greek and Roman texts. The ancient texts were held in high esteem, however, and much of the wrong information about grafting given in these texts was repeated almost verbatim in the new gardening texts.

The Italian Gambattista della Porta published his four-volume *Natural Magic* in 1558. In the third volume, titled *Of the Production of New Plants*, he used information from Greek and Roman scholars, in particular Virgil and Columella, thus repeating the mistakes in compatible stock and scion combinations first stated fifteen centuries earlier. For example, della Porta stated that the fig tree could be incorporated into the plane tree and the mulberry tree, which could also be incorporated into the chestnut tree. Worse yet, he described new misconceptions of what could be achieved by grafting when he claimed that the rootstock could alter the nature of the scion, asserting that mulberry scion on white poplar stock generated white mulberry fruit.

Della Porta was not the only author to exaggerate the uses of grafting. An anonymous author in 1654 credited Dutch and French writers as stating that grafting apple scion on alder or cherry rootstock could make red apples. An alternative method of providing red apples was to soak the scion in pike's blood. Cherries without stones could be produced by using an iron to draw the "heart and marrow from both sides" of the stock cherry tree, anointing it with ox dung, and then grafting another cherry scion onto it. This author also showed a continuing lack of understanding of the limits of graft combinations, stating that apple could be grafted onto apple, pear, cherry, willow, fig, and chestnut.

As scientific observation began to have an influence, some authors, for example, Leonard Mascall

in 1572, showed that they wrote more from first-hand experience and observation, and thus listed fewer incompatible grafting combinations. Mascall was critical of many of his fellow English gardeners for their lack of diligence towards grafting and said that with endeavor there might be many more kinds of fruit flourishing in England. He was critical of claims such as the idea that the scion could take on the nature of the stock: "[M]any which have written that if ye graft the medlar upon the quince tree, they shall be without stones, which is abusive and mockery. For I have (saith he) proved the contrary myself."

The principle of specific fluid

The seventeenth century saw the start of the Enlightenment, or the Age of Reason, which aimed to reform society by reason and the advancement of knowledge through science rather than by tradition, faith, and revelation. Writers of the time followed Mascall's approach to writing from hands-on experience. This was the age of great thinkers and scientists like Voltaire, Adam Smith, Isaac Newton, Robert Hooke, and Robert Boyle. During this time, the English Landscape Garden style was begun by designers William Kent and Charles Bridgeman, and developed by the most famous landscape designer of all, Lancelot "Capability" Brown. While previous designers like André Le Nôtre at Versailles produced formal gardens to show man's control over nature, the English Landscape style embraced nature. This style was in step with the scientific discoveries of the time that showed the wonder of the natural world using the microscope and observation.

An example of the Enlightenment's approach to understanding the world is found in Robert Sharrock (1630–1684), a friend of Robert Boyle to whom he dedicated his book *The History of the Propagation and Improvement of Vegetables by the Concurrence of Art and Nature*. Sharrock was both a churchman, becoming Archdeacon of Winchester, and a botanist. He took a scientific approach to his understanding of horticulture and was particularly scathing of authors who gave magical or romantic explanations of natural phenomena. Through direct experience and observation, he discounted many of the ideas of compatible grafts that had been repeated since the time of Virgil. He was particularly critical of the magical and romantic writing about grafting by people like della Porta. Sharrock stressed the importance of placing the stock and scion together correctly so that the sap could pass between the two plants. He also discussed the limits of compatibility through his own observations. He showed that mulberries would not graft with beech, quince, apple, pear, elm, or poplar as previous writers had claimed. He also showed that mulberries grafted with white poplars would not give white mulberries, that pear grafted onto mulberry would not give red pears, and that the colour of roses could not be changed by grafting.

Although Sharrock took a more scientific approach to his writings on grafting than earlier authors, his explanations still needed to fit in with the accepted understanding of natural processes in the seventeenth century. A theory that had persisted since the time of the Greeks was that of Specific Fluid. According to this theory, every species required a specific fluid to grow and this nourishment was drawn up from the soil though a plant's roots. In the case of a grafted plant, the rootstock could obtain its specific fluid directly from the soil, but the scion would not have this direct contact with the soil and would only get the specific fluid of the rootstock. It would, therefore, be transformed into the rootstock variety.

Pseudo-Hippocrates, writing in the first century BCE, suggested that roots from the scion must grow down through the rootstock and into the earth to get its fluid, providing a possible answer to this problem. Sharrock also had to come up with an explanation of how the scion maintained its identity. It is not known if he was aware of the theory suggested by Hippocrates but, unlike the Greeks, he proposed that the fluid must undergo a "total corruption" when it reached the point where the two plants join. In the case of pear, he suggested that a new sap would arise at the graft point suitable for the pear scion, a process he called "a pear-sap-making power." Although the explanation resembles something della Porta might have written in *Natural Magic*, Sharrock was trying to match up a scientific theory to explain the understanding of the time.

While Sharrock was writing his book, several scientists carried out experiments that would advance the knowledge of plant physiology and finally discount the idea of specific fluids. One of these was Robert Hooke, employed for a time by Sharrock's friend Robert Boyle as his assistant. Born in 1635, Hooke would become the curator of experiments at the Royal Society and surveyor to the City of London after the Great Fire, working closely with Sir Christopher Wren. Hooke was one of the great scientists, not just of the seventeenth century, with wide-ranging interests and achievements. In 1665, he published *Micrographia*, a book describing his microscopic and telescopic observations. It included an illustration of cork cells that he said looked like a monk's cell, thus coining the term *cell* as we know it today.

In the early 1600s, Sir Francis Bacon in England and Jan van Helmont in Belgium conducted experiments on plant growth. Bacon grew roses and other plants in water and concluded that soil was only needed for support. Helmont grew willow seedlings in 90 kilograms of soil and only added rainwater. He grew a 74-kilogram tree and the soil only lost 57 grams in weight. He also concluded that water was the source of growth. In 1699, John Woodward grew spearmint in different sources of water and showed that water with soil added gave better growth than distilled water alone.

It was not until the late eighteenth century that Julius Sachs, among others, used chemical assays to establish that minor soil constituents of nitrogen, phosphate, potassium, and other elements had major importance in plant growth, and accounted for Helmont's weight loss in the soil. It was also in the eighteenth century that Antoine Lavoisier found organic matter was largely formed of carbon and oxygen, and Joseph Priestley and others demonstrated that plant leaves in light take up carbon dioxide and emit equivalent amounts of oxygen. Within 50 years of Sharrock's death, therefore, the idea of specific fluid was shown to be false and ceased to be a paradox in explaining how graft unions formed.

The principle of graft hybridization

From the first writings on grafting, the question of grafting as a form of hybridization has been discussed. While many writers thought that grafting was a form of hybridization, Pseudo-Hippocrates stated that the fruit produced from a grafted plant was true to the scion rather than the rootstock, or a mixture, or hybridization, of the two. The belief that grafting produces a hybrid plant is important in some religions, however.

The Mishna, the Jewish text setting out the laws for people to follow, shows grafting is commonly practised but sets out limitations to its use: "It is unlawful to graft tree on tree, vegetable on vegetable, tree on vegetable and vegetable on tree." What is permissible is to graft plants of the same species and to graft pear onto *krustomal* (a kind of pear). Grafting together different species, such as apple with wild pear, peach with almond, or common jujube (*Ziziphus jujuba*) with Christ's thorn jujube (*Z. spina-christi*) is forbidden. The reason for this and other laws, such as the one forbidding

the wearing of clothes of two different threads, is to keep the Jewish tribe distinct from others. This writing shows that grafting was seen as a kind of breeding, and, like the ban on sowing a field with two kinds of seed or breeding two kinds of animal together, so interspecific grafting was seen as a kind of sexual intercourse that could lead to a new species.

Orthodox Jews still do not allow the grafting of *Citrus medica* (citron or etrog) onto *C. limon* (lemon). Citron is an important part of their Feast of Tabernacles and is a highly prized gift. It was introduced into Palestine in the second century CE, and although it produces little juice, it is rich in essential oils. Lemon started to be used as a rootstock after the seventh century CE when it arrived in the area. In the sixteenth century, the Jewish law against interspecific grafting, or hybridization, was applied to citron, and considerable effort was made to find citron trees that had never been grafted and, therefore, had not been "hybridized." Recent genetic and molecular analysis from citron around the Middle East and North Africa concluded that there had been no genetic introgression between lemon and citron. Thus, the stock may influence the phenotype of the scion, for example, vigour control, but the stock and scion maintain their unique genetic identity.

Even today, some members of the organic agriculture movement question the use of grafting. In 1999, a meeting in Europe discussed the criteria required to evaluate the appropriateness of all available plant breeding and propagation techniques. Three complimentary approaches were agreed to take into account the concepts of "ethical" and "naturalness" in organic farming: the non-chemical approach, the agroecological approach, and the integrity approach. The last approach, integrity, refers to the inherent characteristic nature of plants—their wholeness, completeness, their species-specific characteristics, and their "being in balance" with the organic environment. Being in balance is derived from four different levels: integrity of life, plant-type integrity, genotype integrity, and phenotype integrity. Although the integrity approach is principally applied to breeding techniques, some people question whether it should not also be applied to propagation methods, thus making grafting unacceptable.

Graft hybrids

What is the truth about graft hybridization? Although *Citrus medica* grafts were shown not to hybridize, there are examples of plants in cultiva-tion that arose from grafting and that show the characteristics of both parents, including interme-diate flowers, fruits, or leaves. These plants have been called graft chimeras and also graft hybrids. The term *chimera* comes from Greek mythology and refers to a legendary beast made up of parts of several animals. The term is applied to plants with mutations where the mosaic produced is persistent and cells with two or more different genotypes coexist in a meristem.

One of the most common expressions of chimeras in horticulture is variegated plants where the mutation involved is the loss of the chloroplasts in the mutated tissue, so that part of the plant tissue has no green pigment and no photosynthetic

Citron or etrog (*Citrus medica*), an important part of the Jewish Feast of Tabernacles, using the rich essential oils in its skin. Orthodox Jews still do not allow fruit from grafted plants to be used for this feast.

ability. Chimeras can only be conserved by vegetative propagation and not by seed. They are also commonly unstable and will revert to the original form. *Euonymus fortunei*, for example, has many cultivars with variegated leaves of different formations or colours. In the garden, these forms will often revert to the green-leafed species unless the non-variegated growth is carefully pruned out.

A graft chimera can arise at the point of contact between the rootstock and scion and produces growth with properties of both parents. In 1825, nurseryman Jean Louis Adam grafted *Chamaecytisus purpureus*, normally a low-growing plant, onto a straight stem of *Laburnum anagyroides* to create a small weeping tree. He spotted a sport growing from one of these grafts that had foliage with two types of shoots; the broomlike shoots had three dark green leaflets 1 cm long, and the laburnum-like shoots had lighter green trifoliate leaves 2–3 cm long. When the sport flowered, it produced yellow laburnum flowers on some branches, and dense clusters of purple broom flowers on others. More surprisingly, coppery pink flowers on short (8- to 15-cm-long) racemes, which are midway between the two original grafted plants, were also produced. The leaves on these shoots are also intermediate. This unusual plant proved to be stable if grafted and has given rise to the +*Laburnocytisus* 'Adamii' known in cultivation today. In theory, this graft chimera could

Euonymus fortunei 'Emerald 'n' Gold' showing reversion to the all-green *E. fortunei*. If the green shoots are not pruned out, then the plant will soon lose its variegation.

arise again, but it is believed that all specimens of this plant arise from the one spotted by Adam on his nursery.

Graft chimeras occur rarely but a few others are known in horticulture, like the earlier Bizzarria orange, which arose from a graft of the sour orange and the citron in 1640, and produces fruit of the parents and a compound fruit of the two. Thought to have been lost, Bizzarria was re-discovered in the 1970s by Paolo Galeotti, head gardener of the citrus collection at Villa Medicea di Castello in Florence.

German botanist Hans Winkler first seriously studied graft chimeras in 1907. In his initial experiments, he grafted *Solanum nigrum* (black night-

Two distinct flowers that grow on +*Laburnocytisus* 'Adamii', a graft chimera that is believed to have occurred only once.

shade) onto *S. lycopersicum* (tomato) and then cut back through the graft union. The exposed surface became covered with callus cells of both plants that gave rise to adventitious buds. Most of these developed either pure tomato or pure nightshade growth, but a few showed unusual features. Some, however, were seen to have shoots with one side composed of tomato tissue and the other of nightshade tissue. This is now referred to as a sectorial chimera and is produced when a bud arises from callus at the junction between the two plants.

In later experiments, shoots showing intermediate types of structure were produced, and five types with different leaf shapes and structure were identified. Winkler called these graft hybrids because of the combined characteristics and named the different types as species. Erwin Baur, another German botanist, showed that these were also chimeras and not hybrids and that they arose due to different combinations of cells. *Solanum tubingense*, for example, was shown to consist of a single cell layer of tomato over a nightshade core, while *S. proteus* consisted of two cell layers of tomato over a nightshade core. The cells were therefore distinct as either tomato or nightshade and were not a hybrid to give combined cells. This explains how the intermediate flowers of +*Laburnocytisus* 'Adamii' and the fruits of Bizzarria orange arise: they are not formed by hybridization. Therefore, the stock may influence the phenotype of the scion (for example, vigour control) but the scion and stock retain their individual genetic identity. Mutations occurring at the graft union are therefore graft chimeras and should not be referred to as graft hybrids.

In the Soviet Union of the 1930s, graft hybrids again became an issue with the rise of the theory of graft transformation. Trofim Lysenko (1898–1976), an agronomist, scientist, and communist party worker following the Russian Revolution, promoted Lysenkoism, a form of Lamarckism which had been proposed a hundred years prior by Jean-Baptiste Lamarck (1744–1829). According to Lamarckism, an organism could pass on characteristics that it acquired during its lifetime to its offspring, and, similarly, individuals could lose characteristics they did not require or use. Lysenko rejected Mendelian genetics and its chromosome theory of heredity that was generally accepted by geneticists, postulating instead that hereditary changes to plants could be induced by environmental influences, such as subjecting grain to extreme temperatures or injections. This theory suited the Marxist ideology of Joseph Stalin and led to ideologically driven research influencing Soviet agricultural policy, which in turn has been blamed for crop failures and serious famine.

As far as grafting was concerned, Russian pomologist and fruit breeder Ivan Vladimirovich Michurin (1855–1935) asserted that environmental effects, including grafting, could induce genetic changes. Although this assertion was based on no scientific evidence, it was accepted in the Soviet Union since it fitted with the political ideology of the time. It was not until 1964 that physicist Andrei Sakharov spoke out against Lysenkoism and ended this period of Soviet biological research, and genetics in particular, being politically led.

In 1883, Charles Darwin published *Variation of Animals and Plants under Domestication*. Writing on bud variation in chapter 11, Darwin said, "I will in the first place give all the cases of bud-variations which I have been able to collect, and afterwards show their importance." These bud variations included many examples of bud chimeras, which Darwin called graft hybridizations and suggested were evidence in support of pangenesis, his theory of heredity. According to pangenesis, the transmission of traits is caused by every cell throwing off particles called gemmules, which are the basic units of hereditary transmission. The gemmules were said to have collected in the reproductive cells, thus ensuring that each cell is represented in the germ cells. This theory, which had elements of Lamarckism in suggesting that parents could pass on traits acquired in their lifetime, has been superseded by Mendelian genetics and inheritance. Although Gregor Mendel, an Augustinian monk from Austrian Silesia, had published his research on pea genetics in 1865, it was not widely read until it was rediscovered in 1900, and thus was probably not known to Darwin.

Mentor grafting

Graft hybridization and pangenesis have not gone away, however. In recent years, a technique called mentor grafting has been used, particularly in Japan and China, which is claimed to produce true graft hybrids. One proponent of such hybrids is Yongsheng Liu of the Henan Institute of Science and Technology. He states:

> The most widely adopted mentor-grafting method involving annual plants consists of the following—very young seedlings (from the cotyledonary phase to three- to five-leaved stage) are grafted onto mature stocks (two to three months old, having 20–30 leaves). This ensures that the scions are fully dependent on the mature stock for nutrition. One-way flow of genetic material from stock to scion is affected by removing leaves of the scion (except for two or three at the top) twice a week during the entire time of growth. Usually it has been the progeny seedlings—produced from selfed fruits of the scion, and not the scion itself—that produced various combinations of stock and scion characteristics.

This grafting technique has led to a number of papers published in peer-reviewed journals like

Science, reporting on experiments on tomato, tobacco, soybean, and sweet pepper (among others) that appear to support the theory of graft hybridization. That is to say that, not only did the scion fruit show stocklike phenotypic characteristics, in some cases these characteristics were transmitted to the seedling progeny. These changes included alterations to fruiting direction, fruiting habit, and pericarp colour. Another example of this was in experiments involving grafting cytoplasmic male sterile plants as the rootstock and scions that were not sterile. A significant number of the progeny were found to be male sterile.

How this process of hybridization occurs is unclear and goes against the generally accepted understanding of genetics. Liu suggests that messenger RNA molecules are transferred from the rootstock to the scion. These are then reverse transcribed into cytoplasmic DNA and integrated into the genome of the scion's germ cells. Another author, Yasuo Ohta, proposes that "graft transformation" occurs where chromatin (the combination of DNA and proteins that make up the contents of the nucleus of a cell) is translocated from the stock to the scion. The evidence for these mechanisms is not strong, and the evidence from molecular biology does not provide support for DNA to translocate from one cell to another

Another possible mechanism, this one suggested by Ken Mudge et al. in 2009, is in the recent identification of RNA-mediated gene silencing. It has been found that the silencing signal can occur through the phloem so that gene silencing occurs elsewhere in the plant. This might be an explanation for graft-induced variation and so further research is required in this area.

It should be noted that even the proponents of graft hybrids agree that they can only occur through mentor grafting, and that in all other forms of grafting the scion and stock remain genetically distinct and hybridization does not occur.

2 Uses of Grafting

GRAFTING WAS FIRST USED TO PROPAGATE tree fruit and nut crops where it was the only feasible method of vegetative propagation to maintain selected clones. This function has remained its principal use in horticulture. Selected cultivars of ornamental trees are also genetically heterozygous and must be propagated vegetatively. Here again grafting is used, perhaps because other methods of vegetative propagation are not successful or are too slow or result in poor root systems. Ornamental hardwood trees such as species of *Fagus*, *Sorbus*, and *Betula* are usually grafted as well as cultivars of needle-bearing (whorl-branched) conifers such as *Abies* and *Pinus*. Fewer shrubs are grafted, although many choice shrubs like *Hamamelis* and *Wisteria* are grafted as well as the ever-popular shrub roses.

As developments in rooting cuttings under polythene, mist, or fog using base heat have developed, the use of grafting for many species has declined. The reason for this is that cutting propagation requires less skill and is less expensive than grafting. For example, large-flowered hybrid clematis used to be propagated by a nurse graft. In this type of graft, the scion is attached as low down as possible to the rootstock and only the top half of the scion is tied in, leaving a tail at the base. The graft is potted so that the scion and rootstock are both covered with the growing medium. In this way, the rootstock supports the scion until the scion roots and can support itself. Rooting directly from softwood cuttings has now been successfully developed for these plants and grafting is rarely used. In other species, the choice of propagation method may simply be the grower's preference.

Cuttings and micropropagation are now commonly used to propagate hybrid rhododendrons, as these methods are less expensive to carry out than grafting and they remove the problem of rootstock suckering that can occur with *Rhododendron ponticum* as the rootstock. Grafted plants, however, will grow more quickly and usually flower sooner than those produced from cuttings.

Acer palmatum cultivars are another example of plants that can be propagated by cuttings or grafting. While cuttings, again, will be less expensive than grafting, they can be difficult to overwinter the first year. Stock plants are often forced into growth in greenhouses so that cuttings can be rooted early enough to put on growth before midsummer and to produce enough stores of energy to see them through the winter. Grafted plants do not have this problem and will produce a larger plant more quickly than cuttings will.

Although almost all forestry tree species are grown from seed, superior selections of species (elite trees) are often used to establish seed orchards. Individual trees may show superior characteristics suitable for timber production, like rate of growth and straightness of stem. Seeds could be collected from the parent and grown on, but grafting allows the selected tree to be bulked up quickly and its genotype maintained. Grafted plants will also produce seed much more quickly than seed-raised plants, as grafting shortens the juvenile period in trees significantly. The final benefit is that grafted plants produce seed lower down the stem than seed-raised trees, and therefore seed is easier to collect. For example, at the forestry station of the Sofia University of Forestry, Bulgaria, selected trees of *Pinus sylvestris* (Scots pine) are grafted to increase the number of superior trees for seed production. The grafting also makes the tree mature more quickly and produce cones lower down that can be harvested more easily than a from a seed-raised tree.

The use of grafting, however, has been increasingly used for a range of applications other than propagation. This means that grafting will continue to be important even as techniques improve in other methods of propagation. Other reasons for using grafting follow.

To prevent pest damage

In the 1850s, European grape vines (*Vitis vinifera*), first in England and then spreading on to the continent, began to lose vigour and die. By 1889, wine production in France had fallen to 23.4 million hectolitres from 84.5 million hectolitres just 14 years previously. The plants were seen to have damage to the roots that gradually cut off the flow of nutrients and water to the vine. Secondary fungal diseases also occurred that further weakened the plants.

At first, the cause of this damage could not be found. A microscopic yellow aphidlike insect was often seen near damaged vines, but it was not considered a pest as it was never actually found on the plants. Three noted scientists eventually identified the insect as the pest, and two viticulturists found a solution to the loss of yields in the French vineyards.

35

Grafted *Pinus sylvestris* at the Bulgarian University of Forestry outstation in Yundola, a village in the coniferous forest belt, 1400 metres above sea level, between Rila and Rhodope Mountain. Grafting scion-wood from selected trees not only increases the amount of seed available quickly, but also produces it at a lower height making seed collection easier.

'Phoenix', a variety of grape from Germany used to make white wine, growing in the demonstration vineyard at the Royal Horticultural Society Gardens, Wisley. 'Phoenix' scions have been grafted onto SO4 rootstock. The plant has good resistance to phylloxera, has moderate vigour, and tolerates lime-based soils, but does poorly under drought conditions.

French botanist Jules Emile Planchon is credited with naming *Actinidia chinensis* (kiwi fruit), which was to become such an important fruit crop in the twentieth century. He worked with Pierre-Marie-Alexis Millardet, a French botanist and mycologist, who is perhaps best known for the development of the fungicide made from hydrated lime, copper sulphate, and water known as Bordeaux mixture. Used by gardeners for many years, this fungicide was finally withdrawn from use in the United Kingdom in 2012.

Planchon and Millardet eventually identified the microscopic pest that damaged the roots of grape vines as *Phylloxera vastatrix*, now called *Daktulosphaira vitifoliae* and commonly known as grape phylloxera. The proboscis (the tubular mouthparts) of grape phylloxera has both a canal from which it injects venom into the roots and a feeding tube through which it takes in vine sap and nutrients. As the toxin from the venom corrodes the root structure of the vine, the sap pressure falls and the pest moves on to a new site. Because this insect pest was rarely seen on a damaged plant, as it had already moved on to a healthy root, it took a number of years for it to be identified as the problem. It is a native of North America that had been brought across the Atlantic on plant material in the nineteenth century.

Controversy over whether this insect really was the cause of the problem continued until a third scientist, Charles Valentine Riley, confirmed the diagnosis. Riley became an entomologist for the U.S. Department of Agriculture (USDA), where he was one of the first to practise biological control when he introduced a beetle to control a scale insect that was damaging citrus in California. He confirmed that phylloxera caused damage to European grapes, and he was also the first to note that the American grape (*Vitis labrusca*) was resistant to the pest.

Many methods of control were tried—spraying chemical insecticides, introducing poultry to eat the pest, and burying toads under vines to draw out the "poison" damaging the vines—but to no avail. Two wine growers, Leo Laliman and Gaston Bazille, first tried grafting the European wine varieties onto the American vine. Although this produced plants which were no longer affected by phylloxera, it was not universally accepted as a solution because it was thought that grafted plants produced poorer quality wine than vines grown on their own roots. The effect on quality of wine is still debated. Cyprus is the only part of Europe where grapes are still grown on their own roots, as the island has remained free from the pest. I am not sure if Cyprus is particularly noted as having superior wine to other countries, however.

Other fruit crops with pest issues are often helped by selecting resistant rootstocks. The woolly aphid (*Eriosoma lanigerum*), a pest of apples, can be controlled by using resistant rootstocks. Peach growers in North America used 'Lovell' rootstocks from the 1930s until a root-knot nematode (*Meloidogyne* species) became an increasing problem in California. So 'Nemaguard' was released by the USDA Rootstock Breeding Program in the 1950s followed by 'Nemared' in the 80s and 'Guardian' in the 90s. When a new root-knot nematode species was found in Florida to which the previous rootstocks were not fully tolerant, a new rootstock, 'Flordaguard', was released by the University of Florida.

To increase disease resistance

In the early twentieth century, the Citrus Experimental Station was set up in California when a root rot disease of citrus began to reduce yields, especially in lemon orchards. Above ground, the symptoms are characterized by slow decline, moderate leaf chlorosis, reduced growth, lack of tree vigour and dieback. These symptoms are associated with extensive canker lesions and gummosis at the base of the trunk, as well as root rot, extending from main roots into feeder roots. Gummosis occurs in a number of plants, especially fruit trees, as a reaction to damage caused by weather conditions, infections, insects, or machines. Sap oozes from wounds or cankers forming patches of a gummy substance.

In 1913, plant pathologist Howard Fawcett was appointed to research the causes of these losses. He identified a *Phytophthora* fungus that caused the bark to die near soil level, and caused both gummosis and brown rot of fruits. Several *Phytophthora* species have been identified as causal agents, the most important of which are *P. citrophthora* and *P. nicotianae* var. *parastica*.

Control measures using Bordeaux mixture, copper sprays, and surgical excision of lesions have been used. Further research into rootstocks showed that sour orange (*Citrus aurantium*) was *Phytophthora* resistant, and these were soon being used extensively. However, it is important to ensure the scion is not exposed to the soil or water from the soil, and fungal spray programmes are still used to prevent these problems leading to infection.

Resistant, or tolerant, rootstocks have been identified for many other diseases. For example, although sour orange is gummosis resistant, it is susceptible to the virus tristeza. *Tristeza*, the Portuguese word for "sadness," was the name given to this virus by Brazilian growers, when the diseases caused the death of millions of citrus trees in the 1930s. *Citrus jambhiri* (rough lemon) and *C. sinensis* × *Poncirus trifoliata* (citrange) would be alternative rootstocks for this problem.

Fireblight (*Erwinia amylovora*) is a major threat to apple and pear orchards, and common rootstocks like M9 and MM111 (for apples) and quince (for pear) are susceptible to this disease. An alternative rootstock for apples is M7, and for pears, a rootstock from the Old Home × Farmingdale series. In cherries, Colt or Maheleb rootstocks tolerate bacterial canker (*Pseudomonas* species), but only Maheleb also tolerates crown gall (*Agrobacterium tumefaciens*).

Although grafting is often used to control diseases, it can also cause disease spread. Psorosis (California scaly bark) is an incurable graft-transmissible disease affecting oranges, grapefruit, and tangerine trees. First identified in the twentieth century, bark scaling is caused by a virus leading to death of the inner wood and decline in vigour of the trees that are affected. Psorosis was rare among the seedling trees that made up California's citrus industry in the late nineteenth century. However, as grafted trees became commonly used, psorosis spread and became a major problem by the early twentieth century. Once it was demonstrated that the disease was being transmitted across the graft, measures were introduced to ensure only clean stock was used for grafting.

To control vigour

In some cases, grafting can improve the vigour of the scion. For example, *Sorbus aria* 'Lutescens' is grafted onto *S. intermedia* rather than its own species to give a stronger, more vigorous root system. It is more common, however, to use grafting to control vigour, especially in fruit crops. The use of dwarfing rootstocks for apples has been known for many hundreds of years. 'Paradise', for example, was named in France in the mid-nineteenth century, but is thought to have been used since the time of Alexander the Great. There was also a lot of variation in the size of plants produced by what was meant to be the same rootstock, and, later in the nineteenth century, 14 clones were identified that were all meant to be 'Paradise'.

East Malling Research Station in the United Kingdom (now East Malling Research, or EMR) was established in 1913 to support the fruit industry by researching the cultural problems of growing tree and bush fruits. Captain R. Wellington, the first director and only scientist when the station opened, started to collect selections of apple rootstocks to sort out their classification, test and select the best clones, and standardize those available to commercial growers. This work was just beginning when, in 1914, Wellington left to serve in the First World War.

Wellington's work was taken over by Ronald Hatton, who successfully selected and classified a range of commercial rootstocks. Initially these were listed by Roman numerals; later they were listed by Arabic numerals, all with the prefix M for Malling. Some of these selections were from long-used, named rootstocks, but these were also included in the numbering system. For example, M1 was a vigorous unnamed cultivar, while M2 was the vigorous rootstock previously named 'English Paradise'.

Although ten rootstocks were released by East Malling and have been used worldwide, only two are now used commercially: M7, a semi-dwarf rootstock bred around 1688 in France and previously called "Doucin Reinette" or "Doucin vert," and M9, a dwarfing rootstock selected in 1828 in France as a chance seedling and named "Jaune de Metz" or "Paradis." M9 is still the most important rootstock being widely used around the world.

In 1917, the John Innes Research Station at Merton joined with East Malling to begin a breeding programme to further improve the available rootstocks. The programme's primary goal was to breed for resistance to woolly apple aphids, in addition to their vigour characteristics. In the 1930s the Merton Immune Series was released and given MI numbers (778–793) of which only MI793 is still used. Further breeding, using 'Northern Spy',

a variety that had resistance to woolly apple aphid, led to a second series being released in 1952. This was designated the Malling-Merton (MM) series and numbered MM101–115. Other notable rootstocks have been introduced: M25 in 1952, M26 in 1965, and M27 in 1976. East Malling continues to research into improving rootstocks, using the wide range of genetic material they have collected over the years. MM116, a semivigorous rootstock similar to MM106 but with better resistance to crown and collar rot, was recently released. Selections are being made to improve the rooting of M9, while retaining its dwarfing characteristics to reduce the need for permanent staking or its use as an interstock. Future selections may be for rootstocks more resistant to drought or other climatic factors like windier weather.

The dwarfing rootstock M9 has significantly changed the management of apple orchards, increasing their cropping potential. From the 1950s and 60s, breeding programmes for sweet cherry rootstocks were begun with the aim of emulating the improved orchard management and yields seen in apples. One of these programmes, at East Malling Research, concentrated on crosses between *Prunus avium* and *P. pseudocerasus*.

Prunus avium is the traditional rootstock for sweet cherry, and although some selections have been made to provide improved clonal rootstocks, like F12/1, they produced vigorous trees and were not easy to propagate vegetatively. *Prunus pseudocerasus* is a small flowering cherry that has dwarfing potential, produces root initials, and can be propagated from leafless winter cuttings. Unfortunately, it is not compatible as a rootstock for cherries.

Crosses were made between these two species to try to obtain the benefits of each: compatibility with sweet cherry, reduced vigour of the resulting tree, and the ability to propagate vegetatively the rootstock from cuttings. One cross was eventually selected and released in the 1970s as a new rootstock named Colt. Available from the early 1970s, Colt grows to about 50 percent the height of *Prunus avium* rootstocks, has excellent cropping potential, and can be propagated by leafless hardwood cuttings.

At about the same time in Germany, Werner Gruppe of the Justus Liebig University in Giessen, north of Frankfurt, started a breeding programme for cherry rootstocks with aims similar to those of the East Malling programme. Using *Prunus avium* as one parent, he and his assistant Hanna Schmidt made around six thousand crosses with a range of dwarf cherry species. Out of these, *P. canescens* proved to be the most promising species, and

several selections of seedlings from this cross were made. The two that proved most worthy were named Gisela 5 and Gisela 6 respectively. Although Gisela is a girl's name in Germany, it actually stands for **Gi**essen's **sel**ection for *P. avium* (GISELA).

Gisela 5 was the more dwarfing, about half the tree volume of Colt rootstock, but cropped as heavily. Gisela 6 is as vigorous as Colt, but has twice the cropping potential and is reportedly more regular in bearing. Gisela 5 gives the potential to grow cherries under protection, in a similar way that has been developed for soft fruit in recent years. This opens up new cropping

Sweet cherry trees growing on the dwarfing Gisela 5 rootstock under protection at Castleton Farm near Aberdeen in northeastern Scotland. Without the dwarfing rootstocks making protected production possible, cherries could not be grown commercially this far north.

opportunities and markets for cherries. The only issue with Gisela is that it is not as easy to propagate as is Colt.

The way in which vigour control occurs is still poorly understood, but could be due to the way roots send signals to the top of the plant through hormone and mineral transport through the xylem. Or, it could be the capacity for water uptake and transport through the plant.

Recent experiments showed that resistance to sap flow imposed by the graft union and at the rootstock shanks, was greatest in the highly dwarfing M27 rootstock. In the dwarfing M9, resistance is intermediate and in the semi-dwarfing MM106, it is least. This suggests that in apples, sap flow affects tree vigour, but this may not be the mechanism for all dwarfing fruit rootstocks.

To adapt to soil pH

Rhododendrons are ericaceous plants that normally grow in organic acidic soils of pH 4.2–5.5. Growing these plants in alkali soils requires the creation of raised beds, lined to isolate them from the natural soil, and filled with a low-pH ericaceous medium. Irrigation is by rainwater, and chelated iron (Sequestrene is one brand) should be used if the leaves start to yellow (lime-induced chlorosis). This is all very time consuming and hard work for the average gardener.

Towards the end of the twentieth century, a self-seeded rhododendron was found growing in a quarry with a known alkali soil and bedrock. A consortium of 20 German nurseries crossed this plant with a standard rhododendron rootstock, namely, *Rhododendron* 'Cunningham's White'. The aim was to obtain a new rootstock that would have the lime-tolerant characteristics of the seedling and the compatible grafting ability of 'Cunningham's White'. In addition to these two qualities, seedlings were selected for their habit, vigour,

root development, hardiness, and lack of suckering. The result was the new rootstock 'Inkarho', which grows in soils with a pH up to at least 7.5 and which can withstand cold to –20°C. It also seems to grow well in exposed conditions and even tolerates clay soils.

To survive waterlogging

The arboretum at North Carolina State University has carried out adaptability trials of species from around the world. These have shown that root survival under wet, hot summer conditions is the single most important factor in trees that are not native to the area. Respiration rates rise in hot conditions, requiring more oxygen for the roots. Unique to the southeastern United States is the sudden flooding of poorly drained soils that can create temporary, but fatal, anaerobic conditions for roots. Further west, soil temperatures are high, but soils are drier and oxygen depletion does not occur. Further north, soil temperatures are lower, but rain occurs in cooler conditions, so that anaerobic conditions do not develop.

In the past, rootstocks for grafted ornamentals have been selected only for compatability and availability, but the University now suggests that such rootstocks are not always suitable for climates with wet, hot summers. For example, *Abies* species are used as Christmas trees and so are required in large numbers and at low cost. They are grafted onto *A. balsamea* or *A. fraseri* to keep costs down, but have weak root systems that do not survive in the poorly drained soils of the U.S. Southeast.

Cornus 'Eddie's White Wonder' grows well in the cool, wet Pacific Northwest, but cannot grow in the Southeast. Initially this dogwood cultivar was thought to be unsuitable for the Southeast, but it is now known that the rootstock (*C. nuttallii*) is the problem. *Cornus florida* may be a better rootstock in the Southeast.

To create unique versions of favourite plants

The chrysanthemum is the symbol of autumn in Japan, where it holds a special place in the country's culture. It originated in China and is thought to have been introduced to Japan in the eighth century CE. In the thirteenth century, the retired emperor Gotoba fashioned a sword with a chrysanthemum flower motif. The flower became the symbol of the Japanese imperial household and the Chrysanthemum Throne is the Seat of the Emperor. Chrysanthemum Day occurs on the ninth day of the ninth month, and it was a tradition on this day for people to use cloths to wipe dew from the chrysanthemum on their skin, as a way of maintaining their youth.

Grafted chrysanthemum in shield shape at Naritasan Shinshoji Temple, Narita, Japan.

The chrysanthemum, or kiku in Japan, features in displays in Buddhist temples in the autumn. It is either grown on single stems as a single bloom or grafted so that the resulting plant can be trained into a dome, shield, or different shape. The rootstocks used to give chrysanthemums a strong root system that tolerates high summer temperatures and is disease and pest resistance is *Artemisia annua*, *A. frigida*, or *A. scoparia*. Grafting allows more than one scion to be used so that different colours appear on the shapes produced and it creates stems that are more flexible and thus able to be trained around the frames.

To obtain special forms of plant growth

Producing small trees by grafting onto tall rootstocks has been carried out for many years and was particularly popular in Victorian gardens where standard roses, broom (*Cytisus*), and other shrubs, were widely used. At this time the Kilmarnock willow (*Salix caprea* 'Kilmarnock'), the male form of *S. caprea* 'Pendula', was found and introduced into commerce. In early 1850, a dwarf weeping form of *S. caprea* (goat willow) was found on the banks of the River Ayr in southwestern Scotland, either by an unknown botanist or by Thomas Lang (1816–1896), a nurseryman and seedsman from Kilmarnock, Scotland. Lang named the new form after his hometown. Although the Kimarnock willow could be propagated by cuttings, to obtain the required growth habit it needed to be grafted onto the top of a tall rootstock. A specimen was planted in the Royal Botanic Gardens, Kew, where it was noticed by the horticultural world. It was soon in demand, and by the late nineteenth and early twentieth centuries this hardy, attractive small tree had become (and remains) popular for parks, around water features, and in small gardens.

With the advent of package holidays from the United Kingdom to Spain in the 1970s, the garden has been increasingly used as an outdoor room, and eating al fresco has become popular. Although not new, this led to a demand for patio plants, small standard trees produced from shrubs like *Cotoneaster* and *Euonymus*, grafted in a similar way to the Kilmarnock willow, to enhance the appearance of the patio area.

Cultivation of fruit of the genus *Ribes* began in Europe in the sixteenth century and was introduced to North America by the late eighteenth century. In the nineteenth century, tree culture by grafting *Ribes* was common. In particular, gooseberries were grown by this method, as they can be difficult to propagate by hardwood cuttings, compared with currants like blackcurrants. Raising the canopy by grafting the scion onto stems 80–100 cm tall makes weed control easier, reduces the incidence of powdery mildew, and can be used to mechanize harvesting with over-the-row harvesters. The rootstock used is *R. aureum* 'Brecht' or *R. aureum* 'Pallagi 2'. Both are medium-sized shrubs that have vigorous growth and little suckering.

In the United States, a federal ban on cultivation of *Ribes* species was imposed in the early twentieth century, since the plants serve as an alternative host for white pine blister rust that severely affects *Pinus strobus* (white pine), an economically important forestry tree of the time. Although the ban was removed in 1966, current production of *Ribes* in the United States is still on a small scale and the ban on blackcurrant growing is still in place in many states.

Salix caprea 'Kilmarnock' (left) in late winter just starting to show the catkins.

Tree gooseberries using *Ribes aureum* as the rootstock allow more air movement around a plant than a bush form does, which in turn may reduce mildew and is easier to pick.

Tree redcurrants make an attractive patio plant in their own right.

Tree blackcurrants.

In Europe, especially Hungary, tree gooseberries were popular into the 1920s, but are no longer used for commercial production. Grafted currants and gooseberries are still available for the amateur gardener, however, and are worthy of being used more widely.

The traditional grafting technique recommended for gooseberries is a little unusual and is described as "green grafting." The rootstock is propagated by mound layering, and the graft is carried out in the summer when the rootstock is 80–100 cm tall, still attached to the mother plant, and both stock and scion are about 5 mm in diameter. Grafting when the stock is still actively growing (green grafting) is more successful than waiting until the stock is fully mature and dormant. A splice or whip-and-tongue graft is used and secured with polythene tape to seal the wounds.

To gain the benefits of interstocks

Prunasin is one of several cyanogenic glycosides found in many plants, especially those of the rose family. These glycosides are inactive molecules that are stored in the vacuoles of plants. If the plant is attacked, the glycoside is released and is activated by enzymes in the cytoplasm, producing the toxic hydrogen cyanide (previously known as prussic acid). This makes a valuable defence mechanism against pest attacks. For example, maize is susceptible to damage by rootworms (*Diabrotica* species), while the related sorghum contains glycoside in its roots that makes it resistant to this pest. Prunasin is found in *Prunus* and *Olinia* species and gives the bitter taste to dandelion coffee, a coffee substitute, as well as in *Cydonia oblonga* (quince), which is used as a rootstock for *Pyrus communis* (European pear).

Grafting pear onto quince is an unusual combination, as it is rare for grafts between different genera to be successful. Although a successful union is usually formed, over 20 pear cultivars have been shown to produce delayed incompatibility. That is, the tree may grow for 15–20 years but loses vigour and dies when it should be at its peak of fruit production. Graft incompatibility is discussed further elsewhere in this book, but the solution to this type of incompatibility is to use an interstock between the scion and rootstock.

The reason for pear-quince incompatibility is one of the few that has been fully explained. Pears do not contain prunasin, and once they are grafted onto quince, the prunasin is translocated up to the pear tissue. After the prunasin leaves the plant's vacuoles, it becomes activated by enzymes in the cytoplasm to produce hydrogen cyanide. In compatible grafts, the pear cultivars contain a water-soluble inhibitor of the enzyme that catalyzes the breakdown of prunasin, and so the stage of activating hydrogen cyanide does not occur. Where this inhibitor is lacking and the cyanide is produced, cambial activity is reduced at the graft union, and phloem cells are destroyed around the union.

The activity of prunasin is also temperature dependent, with the optimum temperature being 35°C and no activity occurring over winter. The grafted plant is therefore damaged during the summer, but the cells can repair, and the phloem and cambial connections re-establish during the winter, when no prunasin is produced. There is, therefore, a cycle of weakening and repair that delays the final failure of the graft. In cooler climates, graft failure can be delayed for many years, while in warmer climates graft failure will occur more quickly.

To overcome this incompatibility, an interstock needs to be used that is compatible with both scion and rootstock and thus will contain the inhibitor. The two most commonly used interstocks are Beurre Hardy or Fertility.

M9 is the most common dwarfing rootstock used in apple production in Europe and increasingly in North America. It can easily be kept to 2–3 m high and starts to crop soon after planting, producing large fruit and excellent total yield. It is, however, difficult to propagate and produces fewer shoots in a stool bed than other rootstocks. The root system also develops slowly and the trees must be permanently staked. In addition, although it is resistant to collar rot, it is susceptible to fireblight, woolly apple aphid, crown gall, and mildew. Breeding programmes are trying to overcome some of the disadvantages of this rootstock, while retaining its attributes. Still, M9 can be used as an interstock with good success.

In grafting, the vigour of the root system is important in controlling vigour, but it has been found that M9 still has a dwarfing effect when used as an interstock, especially if long interstock pieces are used. Although the dwarfing effect is reduced when M9 is grafted onto a more vigorous rootstock, it still produces most of the benefits of yield and quality previously described. It is therefore possible to use MM116 as a rootstock with M9 as the interstock. MM116 is a new rootstock that will give the benefits of a stronger root system than M9, which produces a shallow root system requiring to be supported by staking in the orchard. In addition, MM116 will give greater resistance to apple pests and diseases than M9.

Interstocks can also be used to introduce cold or disease resistance. In rubber production in Brazil, for example, *Hevea brasiliensis* (rubber tree) has been affected by leaf blight, but using a resistant interstock can reduce this. In roses, *Rosa rugosa* (hedging rose) was the main rootstock for standard roses, but was affected by disease problems. One solution is to use a rootstock like *R. dumetorum* 'Laxa', normally used for bush roses. An interstock like *R. multiflora* "de la Grifferaie" is budded onto 'Laxa' and grown up to form the stem of the standard rose. The scion of the flowering rose is then budded onto the interstock.

Single stem of *Prunus ×incam* 'Okamé' grafted onto Colt. The new plant has attractive pink flowers in spring.

Prunus serrula grafted onto Colt rootstocks to produce a 1-m stem with *P. ×incam* 'Okamé' grafted on top. The attractive bark of *P. serrula* provides more year-round interest than *P. ×incam* 'Okamé' used by itself.

Yet another use of an interstock is to produce an ornamental tree. *Prunus ×incam* 'Okamé' is an attractive flowering cherry that produces masses of carmine-rose flowers in early to midspring and attractive foliage colour in autumn. Over summer and winter, however, there is not much to recommend it. Inserting an interstock of *P. serrula* (Tibetan cherry) adds a trunk with shining coppery brown, young bark that peels away in bands, giving year-round appeal. The *P. serrula* interstock is grafted onto a Colt rootstock, and a stem is grown up to a metre long. *Prunus ×incam* 'Okamé' is then grafted on top of the stem.

To change existing varieties of trees

Fruit orchards are long-term investments, and although cultivars will be carefully selected, they may not prove to be the best for changing markets, or an orchardist may simple want to grow new cultivars with improved fruit characteristics. This is where top-working may be used to change cultivars in an orchard without grubbing out the existing trees and starting again. Top-working can also be used to graft a pollinator cultivar onto trees that lack proper pollination. It is also possible to graft several cultivars onto one tree as a novelty, although this would not be done for commercial fruit production.

In late winter, select six to ten limbs towards the base of the tree. These should be selected to give a stub of 3–9 cm in diameter once cut back, and form a good framework for the future tree. Branches higher up the tree are retained to provide shade to the new grafts and maintain the growth of the tree during the next season. Either a cleft graft or inlay graft is used, since the scion has a much smaller diameter than the stub. Two to three scions are then grafted onto each stub. After the first year, the successful grafts will have started to grow and all but one should be pruned back on each stub to form the new main branches.

Unworked branches should also be removed after the first year unless the original tree is more than eight years old. In this case, the grafting should be carried out over a few years with about two-thirds of the branches grafted in the first year. The hard pruning of the tree will promote water shoots to be produced. Some of these can be left temporarily to shade the new branches, but should be removed after one season. In this way, cultivars can be changed on an existing tree over a few years. To add a pollinator, only one branch requires to be pruned back and the selected scion grafted onto this.

To repair damaged trees

Bridge-grafting can be used where the bark of a branch or main stem has been damaged, causing the bark to be ringed, preventing the movement of water and nutrients through the phloem and xylem. Inlay grafts are used for this, attaching the scion both above and below the damaged area. The scions themselves are from shoots of the correct diameter from the same tree. This type of grafting is best carried out in the spring just before flowering, and several scions are attached around the stem so that once the union is formed the new growth will be even.

To produce artistic forms via inosculation

Husband and wife, or marriage trees, are recorded in some countries like China, Japan, Georgia, and Scotland. It is believed that branches were deliberately grafted together between trees (inosculation) to symbolize the unification of two people in marriage. At the now-abandoned Lynncraig Farm near Dalry in Ayrshire are two crab apple

trees (*Malus sylvestris*) grafted together in what would have been the farmhouse garden. Another example in Ayrshire is at Eglinton Country Park, where a sessile oak (*Quercus petraea*) on an island in the fishpond appears to have been deliberately grafted. In Thomas Garnett's long-winded *Observations on a Tour Through the Highlands and Part of the Western Isles of Scotland*, published in 1813, a marriage tree was sketched at Inverary in Argyllshire, formed from two stems of a lime tree (*Tilia ×europaea*).

This practice became more elaborate to produce a range of forms by grafting. Called arbosculpture, it is the art and technique of growing stems of trees by bending, pruning, or grafting into a variety of forms. An early exponent of this art form was John Krubsack (1858–1941), a bank president, farmer,

A marriage tree at the now-abandoned Lynncraig Farm near Dalry in Ayrshire. Two crab apples (*Malus sylvestris*) have been grafted together to form a heart shape in what would have been the farmhouse garden.

naturalist, and landscaper from Wisconsin. He also made furniture from wood found around his property, but he decided he could grow a piece of furniture that would be stronger than anything he could make. He eventually grafted trees together to form a seat.

This art form, however, probably reached its peak when, in 1925, Alex Erlandson started to make seats, baskets, and many abstract forms by grafting trees and branches together. For example, his Basket Tree is made from six sycamore trees (*Acer pseudoplatanus*) grafted together with 42 different connections to give a basket shape. Alex decided to make some money from these sculptures and opened Tree Circus, a tourist attraction outside Santa Cruz, California. Although it

In this elaborate form of arbosculpture, stems of *Ficus* species were spiralled around wire hoops until they grafted together to produce a permanent, open shape.

became quite famous both in the United States and abroad, it was not very profitable, and Alex appears only to have made about three hundred dollars a year from his venture. In 1963, he sold his collection of 74 sculptures for twelve thousand dollars. Over the years, these sculptures declined through neglect until the 1970s when Mark Primark, a local architect, began a campaign to save the trees. Eventually Mark received the help of Michael Bonfante who was building a theme park in Gilroy, 40 miles from Santa Cruz. In 1985, the remaining trees were lifted and moved to Gilroy Gardens where 29 of the original sculptures survive.

Although these are extreme examples, arbosculptures are commonly seen. For example, stems of *Ficus benjamina* that have been pleated together are often seen in shopping centres.

To screen plants for disease indexing

Grafting can be used in disease indexing, where latent (unseen) viruses may be present in one plant cultivar that can then infect other cultivars known to be susceptible to the viral disease. The plant thought to contain the latent virus is grafted onto a susceptible plant. If the virus is present, then it will be transmitted to the susceptible plant and symptoms will be seen. The test does not require the formation of a permanent union.

In 1945, a new disease started to infect chrysanthemums growing in greenhouses in the United States and Canada. Within five years, rates of infection were as high as 50–100 percent and threatening the future of the industry. The disease was identified as chrysanthemum stunt virus. Plant pathologists from Cornell University, the USDA, and a commercial company, Yoder, worked to develop techniques that could be used practically to identify the virus. They did this by grafting suspected plants onto indicator plants, and this formed the basis for a certification programme that controls the disease today.

This screening method has now been superseded by dot-blot hybridization (RNA) and a polymerase chain reaction known as PCR. Grafting can still be used to indicate disease problems, however. At present, ash dieback (*Chalara fraxinea*) is causing major losses in this important tree species. Attempts are being made in Europe to breed or select cultivars that are resistant to the disease. In Germany, Heinrich Loesing, director of Versuchs und Beratungsring Baumschulen (Research and Advisory Association for Nurseries), near Hamburg, has chip budded sections from diseased plants onto these new selections to test if they truly are resistant. Unfortunately, many are not proving as resilient to the disease as was hoped.

To micrograft

The only truly novel method of grafting developed in the twentieth century was micrografting, a technique adapted from plant tissue culture (the aseptic culture of cells, tissues, organs, and their components under defined physical and chemical conditions in sterile conditions). In 1902, at the German Academy of Science, Gottlieb Haberlandt presented his experiments in the culture of single plant cells. This led to techniques to study the growth and interaction of cells in controlled conditions. In the 1930s, Roger Gautheret, the French biologist from the Faculty of Science in Paris, produced the first true plant tissue cultures from cambial tissue of sycamore (*Acer pseudoplatanus*).

In the 1950s and 60s, significant progress was made in growing plants *in vitro*. Swedish-born American Folke Skoog progressed the understanding of plant growth regulators, especially

cytokinin. He then took on a doctoral student, Toshio Murashige, to look for an undiscovered growth hormone in tobacco. This search proved fruitless, but the detailed analysis of the constituents of juiced and ashed tobacco led to an improved formula for the medium used for tissue culture. Known as Murashige and Skoog formula, it remains the basis of formula used in plant tissue culture today.

Of more importance to horticulture, this formula meant that tissue culture could move from being just an experimental tool for plant biologists to a practical method of plant propagation. Micropropagation enables the rapid multiplication of plants. For example, new cultivars of rhododendrons, roses, or other shrubs that would

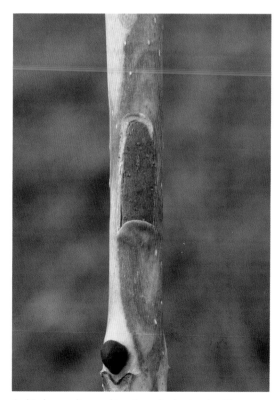

A chip from a diseased tree is grafted onto a healthy plant to test for resistance to ash dieback (*Chalara fraxinea*). Unfortunately, this tree is not resistant.

take ten or more years to bulk up by conventional techniques can be produced in large numbers in one or two years by micropropagation. The technique is also very important in the propagation of orchids that are now such popular houseplants.

One particularly valuable application of micropropagation is in the first stage of producing very healthy plants, especially potatoes, fruit, and other food crops. Viruses and other systemic diseases can usually be eliminated from a plant by the use of thermotherapy and apical meristem culture. By growing the original plant at 30–38°C, disease-free material can usually be obtained by removing the top 0.5 mm of the meristem and culturing this *in vitro*. With fruit like strawberries and raspberries, these shoots can then be multiplied in culture, and once enough plants have been grown, they can be induced to root and weaned back into natural growing conditions.

Micrografting has been developed for citrus plants because the scion material will not produce its own roots. In this case, seedlings of rootstocks are grown *in vitro* and the small shoot tip of the scion material is then removed to be grafted onto the rootstock. A triangular or T cut is made on the top of the seedling, and the scion shoot tip is placed on top. The procedure is carried out on a sterile bench, called a lamina flow cabinet, to prevent contamination, and then the graft is placed back into the culture tube. Once the micrograft has started to grow and it is known to be free of viruses, 1- to 2-cm-long shoots of the scion are re-grafted onto conventionally grown seedlings that are grown on in natural conditions. This very specialized procedure has proved invaluable to the citrus industry.

There have been other applications of micrografting used recently. Forestry has micrografted conifers, especially old trees that have particular value. Micrografting is being used to propagate these plants vegetatively to increase their number and return them to a more juvenile stage. It is hoped that this material can then be propagated by cuttings to provide plants for a seed orchard. Micrografting is also used by research scientists to investigate aspects of plant physiology like graft incompatibility.

3 Formation of Graft Union

THE NATURAL WOUND RESPONSE IN TREES and other perennial plants begins as soon as they are damaged. Grafting is the manipulation of this process to induce a permanent connection between the rootstock and scion. Grafting only has a chance of being successful if the scion and rootstock are compatible. The quality of the plants for the selected combination must also be suitable in trueness to type, girth of shoots, root development, and lack of pests and disease. The grafter's skill is required to produce the correct cuts to enable a close vascular cambium connection between the plants. Finally, the correct environmental conditions are needed to prevent the desiccation of the scion while it is not connected to a root system, and to induce the graft union to form.

Natural response to damage in wood

Alex Shigo, former chief scientist with the USDA Forestry Service, described the model for the compartmentalization of decay in trees (CODIT) in the 1970s. According to this model, trees and other woody perennial plants do not repair damage but seal it off to prevent the spread of further damage or disease. A cross section of a tree reveals how highly compartmentalized it is (see Figure 3-1).

Bark consists of both an outer layer and an inner layer. The periderm, or outer bark, consists of cork on the outside, cork cambium in the middle, and phelloderm on the inside. This outer bark is important for the prevention of water loss and the insulation of internal tissue. Next to the periderm is the inner bark, or phloem, which is made up of several living cell types including sieve elements, companion cells, fibres, and parenchyma.

These cells move sugars from the leaves to other plant parts.

The vascular cambium, which comes between the bark and xylem, consists of meristematic cells that divide and differentiate to produce new xylem and phloem cells that lead to an increase in the tree's girth. Xylem is principally wood and is made up of live and dead cells that have walls of mostly cellulose and lignin. Wood that is living is called sapwood and has three functions: to transport water and nutrients from the roots, to store energy reserves, and to provide mechanical support. Wood made up of cells with no living contents is called heartwood. It provides the main mechanical support to the tree and may be a repository for waste metabolic products (polyphenolics or tannins).

Together, the phloem and the xylem form a continuous system of vascular tissue extending throughout the plant. Some plants also contain pith in the wood, soft spongy parenchyma cells

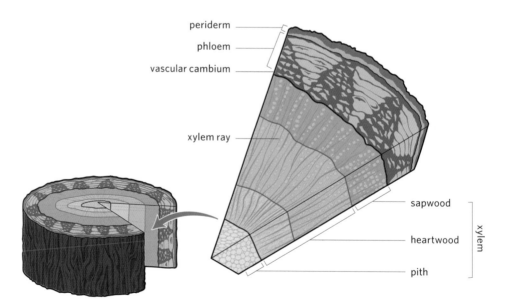

Figure 3-1. Diagrammatic representation of the structure of wood.

that store and transport nutrients. These structures divide the plant into compartments, in particular the growth rings that go round the girth of the stem and the xylem rays that extend out from the centre.

When wounded, perennial woody plants are able to isolate compartments to contain an injury and prevent disease from entering and spreading from the damaged area. Damage is, therefore, not actually repaired but contained and sealed off. Four walls are produced around a damaged area to isolate the damage. First, the vertical vascular cells are plugged; this is known as Wall 1. Wall 2 develops to prevent the inward spread of damage. Wall 3 prevents the lateral spread around the stem. Lastly, Wall 4 resists spread into newly formed xylem or wood that grows after the time the tree was wounded. The walls are not visible and are separate from the visible wound closure. In wound closure, undamaged parenchyma cells divide rapidly to produce callus tissue (the white undifferentiated cells that can be seen to form around a cut surface). This callus will form the wound periderm that eventually covers the cut surface.

Woody plants may be weak or strong wound compartmentalizers. Wall 2 is the most important in preventing inward spread of microorganism-caused discolouration towards the centre of the tree. This discolouration may extend to the pith in weak trees. Although the majority of successfully grafted trees are strong compartmentalizers, for example, *Castania sativa*, *Carpinus*, *Gleditsia*, *Juglans*, *Pinus rigida*, and *Robinia pseudoacaia*, many commonly grafted trees are weak compartmentalizers, for example, *Aesculus*, *Betula*, *Cercis*, *Fagus*, *Malus*, *Prunus*, and *Salix*. How easy or difficult a plant is to graft may be more closely linked to the amount of callus produced to form the wound periderm. In the strong compartmentalizer *Juglans* and the weak compartmentalizer

Aesculus, little callus is produced and they are difficult species to graft.

The process of the graft union initially depends on the natural response to damage in woody plants. A necrotic, or dead, layer of cells forms several layers thick when a stem is cut. This isolates the damaged area and the process of wound compartmentalization begins. In grafting, by making straight cuts to enable close contact between the scion and stock and by applying pressure at the cuts, a thin necrotic layer is produced that starts the cohesion between the rootstock and scion. If no pressure is applied, if uneven cuts are made, or if the cells are overly turgid (as when cuts are made immediately after rain), then a thick, irregular layer of necrotic cells may be produced.

Callus formation will then begin over the cut surfaces. Any living cell has the potential to form callus as long as the cells do not lack turgor. Moisture is retained around the cut surfaces by sealing the cuts with wax or polythene tape or, where this is not possible, covering with a plastic tent to retain a high humidity. If moisture is not retained, then the lack of turgor prevents cell expansion and callus formation, and thus the formation of the new vascular cambium connection. It is, therefore, very important to manage water relations correctly within grafted plants.

Before the vascular cambium connects between the scion and rootstock, the callus formation allows water to pass from the rootstock to undamaged conducting tissues of the scion. This is required for the survival of the scion at this time. Callus is at first produced at random, with the cells dividing to fill the gaps between the two cut surfaces. This is another reason why even, straight cuts are important; poorly made cuts require more callus to fill the gaps, which in turn slows down the formation of the graft union, if it is able to form at all. The production of the mass of callus cells also forces the scion and rootstock apart unless tied correctly. It also breaks up and digests the necrotic layer of cells, thus preventing them from forming a barrier to the graft union.

Adhesion between the cuts further develops by beadlike materials of pectins, carbohydrates, and proteins that project from the surface of callus cells and glue the cells from the scion and stock together. This process has sometimes been called an organic weld as these materials join interlinking parenchyma cells from the two different plants to form a single entity, although they retain their two separate characteristics. At this stage, cell division is becoming more ordered, and plasmodesmata (microscopic channels allowing transport and communication between cells) form anew. Rows of cells form at right angles to the pressure exerted, aligning the rows across the cuts.

The next stage of the graft union is the formation of a new vascular cambium connection. The time taken for cambium cells to appear after the callus formation depends on the species as well as temperature and quality of the graft preparation. Given optimum conditions, it takes 7 to 21 days for cambium cells to form in *Malus*, 22 days in *Pyrus*, 21 to 42 days in *Picea abies* and *Pinus sylvestris*, and up to 90 days in *Picea sitchensis* and *Populus*.

Cells close to the existing cambiums of both the rootstock and the scion develop first and spread until they join the two cambiums together. Once this connection has been established, the new cambium produces normal secondary tissues of the phloem and xylem. These processes involve the formation of sieve plates in the phloem that transport metabolites from the scion to the roots and perforation plates in the xylem, enabling the movement of water and nutrients from the roots to the scion. Only after this happens can the new growth occur in the grafted plant. It is, therefore,

important that the scion buds are dormant during this process.

Cambial union will still occur if the cambiums are not perfectly aligned when tying a graft together. In fact, even if they are well aligned, the cambiums can be pushed apart by callus formation without detrimental effect on the final union. Only when lateral alignment is very poor will the cambiums not connect and the graft will fail. For example, if the cambiums fail to align when the scion is much thinner than the rootstock, it is better to align the scion with one side of the rootstock. This will allow a union to form on one side of the graft successfully, while a groove develops down the other side of the union as the secondary tissues fail to form. It is best that the scion should match the diameter of the rootstock. If the two parts cannot be matched up, then a modified graft technique can be used, such as the chisel graft (see page 121).

Factors affecting successful graft union

Manipulating natural processes is not the only requirement for a successful graft union to occur. Selecting a suitable combination for the graft is important. Likewise, creating the correct physiological and environmental conditions also affects the success or failure of a graft union. Grafts have to be carried out at the correct time. Normally this is when the scion buds are dormant. There must also be some cell division occurring in the rootstock, which is affected by the temperature during grafting. Another condition for successful grafting is making a close cambial connection by holding the two plants together with the correct pressure. Finally, it is important to maintain sufficient humidity and temperature for the graft to form and to prevent desiccation of the scion.

Compatibility

For a graft to be successful, the scion and rootstock must be compatible (able to form a permanent union). Unfortunately, incompatible scion and rootstock may still successfully form a graft union that will grow for a number of years before the union fails.

Graft failure may occur if poor-quality plant material is used, if cuts are poorly made or matched up, if ties are poor, if the post-grafting environment prevents the union from forming, or if the two plants are incompatible. Signs that a graft combination may be incompatible include a high proportion of grafts failing to form a union, poor vigour, yellowing foliage, early defoliation, shoot dieback, uneven growth in the diameter of the scion and rootstock, differences in the growing season of the scion and rootstock, suckering, and early death of the grafted plant. All these symptoms, however, may be due to other factors like soil conditions or disease.

The one definitive symptom of graft incompatibility is if the plant breaks cleanly at the graft union. However, if a number of symptoms arise, then incompatibility must be suspected. Immediate graft failure will often occur between an unsuitable stock and scion. *Acer rubrum* cultivars, for example, grafted onto *A. platanoides* quickly deteriorate shortly after bud break.

Unfortunately, incompatible combinations will often form what appears to be a good union and the scion will start to grow. When a graft fails over a long period, defined as after one to five years, it is referred to as partially delayed incompatibility. When *Pyrus calleryana* 'Chanticleer' was grafted onto Quince A, a rootstock used for culinary pears, the graft grew well in spring but stopped in midsummer, and yellow foliage appeared in late summer. The plant did start to grow again the next spring but, on examination, it was seen that minimum hand pressure

would break the union and the graft would fail completely after a few years.

Delayed incompatibility is used to describe a combination that fails after 15–20 years. This occurs, for example, when 'Bartlett' pear is grafted onto quince.

In general, there is a genetic basis to combinations being successful. Incompatibility may occur due to physiological differences between the scion and rootstock, or abnormalities developing across the graft union in the vascular cambium. It is likely that the more closely related the graft combinations are (for example, when cultivars are grafted onto rootstock of the same species), then the greater likelihood of success. Grafting a

The graft union on this beech tree can still be clearly seen as a distinct line, suggesting that the plant probably has a delayed incompatibility problem and will fail when it is still comparatively young.

cultivar onto a seed-raised rootstock of the same species usually produces a successful graft union. However, even here some incompatible combinations can occur, as is the case in *Quercus rubra*, *Acer rubrum*, and *Castanea mollissima*. Combinations that are more distantly related are less likely to be compatible.

A number of interspecific crosses are commonly used. For example, almond (*Prunus dulcis*) is grafted onto rootstocks of peach (*P. persica*). A few intergeneric grafts are also important; many pear cultivars (*Pyrus communis*) have long been grafted onto quince (*Cydonia oblonga*) to produce dwarf trees. Where delayed incompatibility occurs between species or genera, then nurse grafts are sometimes used. *Ligustrum ovalifolium*, for example, can be used as a rootstock for *Syringa vulgaris* cultivars. Such a combination will give delayed incompatibility, but as long as the graft is made as low down as possible and is planted so that the union is covered, then the graft will support the scion long enough for it to produce its own roots. There are no examples of successful grafts between families.

Genetic differences cause graft failure for several reasons. In genus *Prunus*, incompatible cherry combinations fail to develop the phloem sieve tubes properly below the union, preventing the movement of photosynthates across the graft. In apricot and plum combinations, the vascular cambium tissues do not fully form and a weak union is made that soon fails. In incompatible apple grafts (genus *Malus*), it appears that the parenchyma cells interrupt the xylem when the vascular cambium does not develop correctly.

In some cases, incompatible combinations can be successfully grafted by the use of an interstock that is compatible with both the scion and rootstock. This is called localized or non-translocatable incompatibility. In other cases, the use of a mutually compatible interstock does not prevent incompatibility. This is called translocatable incompatibility and the well-known example is 'Hale's Early' peach grafted onto 'Myrobalan B' plum rootstock. Even when a mutually compatible plum interstock ('Brompton') is used, the graft still fails. An accumulation of carbohydrates occurs in either the scion or the interstock due to the breakdown of the phloem, possibly due to the movement of organic molecules.

But what about combinations of the same species where genetic differences are minimal? Frank Santamour of the USDA has suggested that failure in these instances is due to differences in the cambial peroxidase enzymes that occur in lignin. These have a role in a plant's defence against pathogens and may also have a role in the formation of lignin, an important component of secondary walls in plants that strengthens the wood (xylem) of trees. Lignin is a complex organic polymer and is unusual because it lacks a primary structure and is, therefore, very heterozygous in its nature.

Peroxidase has several isozymes (enzymes that differ in amino acid sequence but catalyse the same chemical reaction). In genera with compatibility problems within a species, several isozymes of peroxidase can be present in different combinations. In particular, three major isozymes seem to be important. It is suggested that if the scion and rootstock do not have the same combination of these isozymes, then incompatibility is likely to occur.

The genus *Quercus* has particular incompatibility problems even with scion and rootstock of the same species. In the United States, native *Q. palustris* (pin oak) has potential as a street tree because it transplants well, grows over a large geographical range, and has attractive autumn colour. Unfortunately, because it has a drooping growth habit that requires continual pruning to keep it above the tops of vehicles and pedestrians, it is not used. More-upright cultivars were identified, but have not become widely available due to delayed incompatibility problems that appear in the nursery several years after the graft is made or even planted in the landscape. Oaks are divided into two subgenera: white oaks (for example, *Q. robur*) and red oaks (for example, *Q. rubra*, *Q. palustris*). Red oaks have a large range of isozyme types and thus grafts in this group have incompatibility problems. White oaks have less isozyme variability and therefore fewer incompatibility problems.

Incompatibility between *Acer rubrum* cultivars and *A. rubrum* rootstocks has led to the development of cutting propagation as an alternative method of producing cultivars such as *A. rubrum* 'Schlesingerii'. Unfortunately, at present, grafting is the only feasible vegetative method of propagation for the oaks.

Another cause of incompatibility may be the presence of virus or phytoplasma in the scion or stock. Phytoplasma are pathogens of plants most often found in tropical and sub-tropical species. Like viruses, phytoplasma are transmitted by vectors such as aphids, but unlike viruses, phytoplasma are bacteria-like organisms. They cause a range of symptoms in plants from mild effects like yellowing of leaves to the death of the plant. Because they are translocated in the phloem, they can move across a graft union.

The virus tristeza is a major disease problem in citrus. Initially, the failure of *Citrus ×sinensis* (sweet orange) grafted onto *Poncirus trifoliata* (Japanese bitter orange) was thought to be due to incompatibility, and alternative rootstocks like *Citrus sinensis × Poncirus trifoliata* (citrange) were used successfully. However, it was noted that the problem initially occurred only in South Africa and Java, while in other countries the sweet and sour combination was successful. Research finally identified a virus that was causing this disease. It is tolerated by the sweet orange and so could be

present in the scion and show no symptoms. The sour orange, however, is susceptible to this virus which was translocated across the union following grafting, causing the death of the rootstock. An alternative to selecting another tolerant citrus rootstock to the virus would be to clean up the virus from the scion.

Failure of grafts due to viruses is also found in other species. In *Juglans regia* (walnut), walnut blackline is caused by the cherry leafroll virus and transmitted by pollen. In apples, apple union necrosis and decline is caused by tomato ringspot virus and transmitted to the rootstock by

Early autumn colour on the oak in the centre is a sign of incompatibility between rootstock and scion. The young plant will be removed from the crop. Unfortunately, other plants in this crop may have delayed incompatibility and will fail after the trees have left the nursery.

nematodes. Some combinations such as Golden Delicious on MM106 are particularly susceptible to this virus. In *Pyrus*, pear decline has been found to be due to a phytoplasma.

Compatibility, therefore, remains a major problem for propagators. Since there is no single explanation for how incompatibility occurs, or a single cure, and since it can be delayed for many years, introducing a new graft combination is fraught with dangers. Some researchers have attempted to develop methods for predicting incompatibility. Santamour has proposed isozyme separation by electrophoresis as a possible method of predicting where variation in cambial peroxidase enzymes affects compatibility. When isozyme phenotypes of stock and scion are found to be similar, then a compatible union is possible, since the same type of lignin will be produced. When the isozyme phenotypes are different, then vascular continuity will not be re-established between stock and scion.

The similarity of proteins between stock and scion has also been suggested as a means of predicting compatibility. Several methods of comparing proteins have been tried and may offer future methods of prediction. *In vitro* techniques to help propagators avoid compatibility problems are also being investigated. In practice, testing new combinations for compatibility is expensive and time consuming. In the meantime, publications providing suitable stock and scion combinations should be consulted when grafting an unfamiliar species.

Timing

In general, grafting is carried out when the scion-wood is ripe, the buds are dormant, and the rootstock shows signs of at least some active growth. For temperate plants, this means that grafting is mainly carried out in early spring or late summer.

The precise timing usually comes down to the choice of the propagator. *Acer palmatum* can be bench grafted over a long period because it only puts on one flush of growth; it can be grafted from early summer to autumn and also in the early spring. Many other plants can be grafted in the early spring as well as the late summer, as long as heat can be provided correctly to the graft. *Wisteria* and *Abies* species are grafted in late winter to early spring. Often the timing depends on when the nursery has time to do the grafting, or when it has had success with the species. Some propagators of *Fagus sylvatica* only find success in early spring with heat, while others find success in late summer.

The ripeness of the scion-wood refers to its firmness. Soft, flexible stems are under ripe, while over firm, older stems are often overripe. Ripeness is due to the carbohydrate and nitrogen ratio (C:N) within the wood. In spring, when the new shoots are growing, all the energy produced by photosynthesis is used for growth, and there is a low C:N ratio. Once extension growth stops, the energy from photosynthesis starts to be stored as carbohydrate (sugars) within the stem. Lignification occurs and the stem becomes firm, or ripe. At the same time, the buds are maturing.

Carbohydrates give winter hardiness to a stem and provide reserves of energy that get the plant through the winter and enable the buds to grow in the spring. In vigorous plants such as willows (*Salix*), dieback often occurs over winter at the shoot tips because the tips have not properly ripened and had time to store carbohydrates. Instead, they remain soft and prone to damage by the cold.

Generally, scion-wood is up to one year old. As wood becomes older or thicker, the ratio of C:N increases and is not suitable to use for grafting. As well as age, the C:N ratio is affected by the nutrition of the parent plant. It is important, for example, not to apply too high rates of nitrogen to the parent plant to avoid the growth remaining too soft and unsuitable for grafting.

In practice, selection of suitable scion material comes with experience. It has to be firm but not overly rigid. In theory, sugar levels within the stem could be measured for their degrees of Brix. One degree Brix is 1 gram of sugar in 100 grams of solution, and is used in the wine, sugar, fruit, and honey industries. As yet, the required levels of degrees of Brix have not been identified, and it is debatable if the cost of this work would make a significant improvement in graft success.

In late summer, new buds remain dormant as day length is decreasing and temperatures, especially at night, are reducing. It is, therefore, relatively straightforward to ensure that the buds remain dormant while the graft union forms. In early spring, the scion buds will start to become active. Grafting has, therefore, to be carried out early enough so that the graft union will form before the new shoots develop from the scion. Grafting early to ensure the scion bud is dormant may mean that the rootstock is not in a suitable condition for the graft union to form, as it has to have started active growth. Where possible, it is best to collect scion material in midwinter when it is fully dormant. Then cold-store (refrigerate) it so that the buds remain fully dormant and the graft can be carried out in late winter or early spring when the rootstock is ready.

The type of graft used may also determine the time of year it is carried out. T-budding of roses must be carried out in summer once the bark is slipping (that is, when the bark separates easily from the wood so that the scion bud can be inserted behind the bark). Chip budding, which has largely replaced T-budding for deciduous trees, is also mainly carried out in the summer. Since it does not require the bark to slip, however, it can be successfully carried out over a longer period in the summer. It is also possible to use this method in winter.

The condition of the rootstock is critical to successful grafting. It has to be free from pests and diseases, be of the correct diameter, and be in active growth without excessive sap flowing through the stem. In spring, the sap should be starting to rise and the rootstock should be starting into growth so that there is cell division for the graft union to form. However, there should not be excessive sap that could push the grafts apart due to the pressure. In the summer, extension growth is ending and the flow of sap is reducing as the dormant season approaches. Again, the sap needs to be controlled, so that there is still active growth but the sap is not excessive.

Field budding techniques attach a bud to the side of the rootstock that still has the top growth attached. This acts as a sap draw to prevent a build up of sap and pressure at the graft union so that budding can be carried out in the summer as soon as the scion buds are mature enough, and, if T-budding, the bark slips from the wood. In bench grafting, side veneer types of grafts can be used that also keep the tops of the rootstock in place to act as a sap draw while the graft union forms. This type of graft can be used in spring or summer.

Grafts that do not use a sap draw, but simply cut off the top of the rootstock and replace it with a bud-stick from the scion—whip types of graft—are an alternative for bench grafting, apart from conifers. These grafts require precise management of the rootstocks to control sap flow, but, if correctly managed, the grafts and aftercare are generally straightforward compared with the side veneer. Although grafts that do not use a sap draw are normally carried out quite late in the season, they can be carried out in early summer in some species that only put on one flush of growth in a season. *Acer palmatum* is one example; once its single flush of growth is mature enough, it can be grafted as long as sap flow is managed.

In areas of the United States that have long growing seasons, budding can be carried out in early summer. The aftercare of these grafts is critical, since the scion buds will grow actively soon after insertion. Early summer budding is used by nurseries to get scion growth in the first year and to reduce the time a tree spends in production before it is sold.

Timing may also be affected by the weather conditions at the time of grafting. Grafting records kept in Australia over 12-years have attributed over 99 percent of major losses to budding or grafting during wet weather. It is suggested that fully turgid tissue collected during wet conditions bruises five to ten cells into the cambium when cut and prevents compartmentalization. Healthy callus tissue does not grow and bacterial soft rots develop.

To avoid this problem, it is recommended to collect scion material in midmorning or afternoon, or 24 hours after the last rain, and that it is never grafted onto wet rootstocks. With bench-grafted material, some propagators also ensure that grafted plants are not watered overhead for the first three months to stop water getting into the graft. Studies of *Acer platanoides* 'Crimson King' over several years showed that once the bud had taken, there was no link between bud take in the field and whether the growing season had been wet or dry.

Quality

In the 1960s, experimental work showed that over twenty viruses occurred in apple and pear trees. In some varieties, these viruses show symptoms like leaf spot or malformed fruit. Apple mosaic virus, for example, causes bright yellow patches on the leaves of Golden Delicious and Lord Lambourne, and affects the growth and yield of the trees. In these cases, infected trees can be removed from an orchard. In Cox, however, this same virus causes no obvious symptoms and the infection is considered latent. When virus-free trees of Cox were produced, however, it was shown that the latent

virus still caused a yield reduction of at least 20 percent. This example shows the effect of just one virus. The reduction in yield has been shown to be cumulative if more than one virus is present. Therefore, a need for an improvement in the quality of propagation material was identified for economically important fruit trees.

Techniques were developed where plant material was treated to remove viruses and other disease-causing organisms. These were then grown on in isolated conditions to reduce the chance of virus transmission and tested to ensure that they were true-to-type and disease free. Cultivars of rootstocks and scions that met the health requirements were given a Plant Health Certificate and could then be sold to nurseries. This process ensured that propagators of fruit trees were starting their production with healthy plant material, and enabled orchards to improve their yield and fruit quality. Because the programme is expensive, only economically important fruit crops were put through this process. For ornamental species, where certified stock is less likely to be available, propagation material should be grown on designated stock plants or bought in from a reputable supplier.

Seed overall does not transmit viruses, and so seed-raised rootstocks tend not to have the same issues with viruses. Nevertheless, *Prunus avium* seed should be virus tested when using it as a rootstock for ornamental cherries.

Another important consideration when obtaining quality rootstock material for woody plants is the stem diameter. *Acer platanoides* 'Crimson King', for example, has been studied because of its difficulty and variability to graft. Trials were carried out on rootstocks within the 6–10 mm range, and stock a year older that was thicker than 10 mm. The first-year stock gave an average take of 76 percent compared with a 26 percent take on the second-year larger-diameter stem.

In all types of grafting, the quality of roots is very important. Further trials with 'Crimson King' showed that the growth of larger structured root systems improved the graft take significantly. In bench grafting, the roots need to be active at the time of grafting. It is recommended that there be about 6.5 mm of fresh white root prior to grafting onto pot-grown rootstocks.

It is possible to bench graft many species using either bareroot or pot-grown rootstocks. Bareroot plants can be used in the spring for deciduous plants, and have the advantage of being less expensive to produce than growing the rootstock in a container for a year. In some cases, the successful take may be less with bareroot plants. For example, bareroot *Magnolia* has been found to give about 20 percent less take than if the rootstock is container-grown. For most species, however, the take may be about the same, but the subsequent growth may be more varied with the bareroot material, which in turn can affect the number of plants that subsequently make the required grade.

Alignment and tying of stock and scion

As discussed earlier, the vascular cambiums of the rootstock and scion do not need to be in contact for a graft union to form. However, the quicker the union forms and the larger the aligned area is between rootstock and scion, the stronger the subsequent growth. With *Juglans*, *Aesculus*, and other genera that are difficult to graft because they produce little callus, close cambial contact is critical. With easier-to-graft genera such as *Malus* and *Sorbus*, however, there is more tolerance in the alignment, and the callus bridge will form even if the scion and the stock are slightly misaligned. Ideally, the diameter of the rootstock and that of the scion should be closely matched, and long, clean, straight cuts should be made to give as large a surface area in contact as possible.

The skill of the propagator is important in ensuring that the cuts of the stock and scion align. Straight, clean cuts ensure no gaps between scion and rootstock once the graft is complete. In addition, the better the carpentry of the grafter, the quicker the graft union will form and reduce the chance of graft failure or the entry of disease.

The value of aligning the cambiums of stock and scion was illustrated by work on budding carried out at East Malling. Chip budding not only proved to give higher percentage of success in fruit and ornamental trees than T-budding, but the length of maiden growth in the year after budding was greater, and more laterals were produced in the second growing season. The greater success occurred because of faster take in chip budded material due to the closer cambial contact than in T-budding, where the bud is placed behind the bark. Some results also showed chip budded plants to have more frost resistance than T-budded. These results reported in the 1970s have led to chip budding largely replacing T-budding as far as tree production is concerned.

It is important to tie the graft tightly together so that the vascular cambiums are closely aligned and held secure. Any movement at the graft due to wind, or from handling, while the union is forming, is likely to disrupt the callus bridge development or misalign the cambiums so that the connection cannot form. A tight graft union will also help to pull the cut surfaces together if less than perfect cuts have been made, so that the cambiums are in close proximity and the possibility of desiccation of the scion is reduced. To achieve a close union, the stock and scion are tied together, typically with polythene tape, parafilm, or rubber strips, although other materials have been used like raffia, cassette tape, and even super glue.

In some instances, the type of tie used may be left to the preference of the grafter. Rubber ties have the advantage of being biodegradable and usually do not need to be removed by the propagator. However, with softwoods, like conifers, and with some hardwoods, like birch (*Betula*), the rubber tie often does not degrade quickly enough and should be removed once the graft union is secure to prevent choking or cutting into the stem. Rubber ties can actually kill a high proportion of birch if not removed in time. In most cases, however, rubber goes brittle in sunlight so that when the new plant starts to grow and the stem swells, the rubber breaks apart. With other plants, rubber ties can degrade too quickly and slacken off before the graft has taken, thus exposing the graft to easy damage by any movement.

If a rubber tie is twisted around a graft, it can cut into the expanding stem and cause a corkscrew effect, so for this reason, propagators may prefer using polythene ties. These do not biodegrade, so the propagator has an extra job to do in removing them. But, removal of the tie is fully in the control of the propagator and so can be done once the union is strong enough but before any constriction of the new stem may occur. The callus at the graft, which initially appears white and then colours green, will turn brown when it is time to remove the tie. One type of grafting tape known as Buddy Tape is made from olefin resins and has become popular in recent years. It is meant to be biodegradable but in practice may not break down soon enough, especially with winter grafting, and may still need to be removed by the propagator, especially if wrapped more than two or three times round a graft.

Some types of graft apply pressure at the union and hold the graft more firmly together. A straight whip (splice) graft is a quick and successful technique for many deciduous, bench-grafted trees. Some grafters prefer to make a whip-and-tongue graft where an extra cut allows the stock and scion to be wedged together. This technique may be particularly beneficial with *Aesculus* where even the

slightest movement at the graft union will probably cause it to fail, but it is not recommended for *Juglans regia*, another species which is difficult to graft. Because *Juglans* has a pithy stem, a whip-and-tongue graft causes more damage to the cells from the tongue than a whip graft, and produces less than desirable results.

It should be noted, however, that once the grafts have taken, there is no difference in subsequent growth whether the whip or whip-and-tongue method was used. Tying the grafts firmly together reduces the chance of gaps between the cut surfaces that could lead to the desiccation of the scion. The pressure applied at the graft union has also been shown to be important in the process of callus cells differentiating to form the cambium, xylem, and phloem tissues, and ensuring these are correctly aligned. That is, the cells forming the callus bridge need to be induced to form the specific cells of the cambium, xylem, and phloem tissues that will properly align to form the connection across the graft.

Prevention of desiccation

Since the scion graft wood, or bud, is not attached to a root system, it can quickly lose water. In summer, graft wood should be collected when it is fully turgid, that is, when the cells are fully hydrated. This is best done in the morning. To prevent further desiccation, the leaves should be removed from the bud-sticks as soon as possible, and, once the cuts are made to form the graft union, they should be tied quickly to avoid loss of water.

Once tied together, inhibition of callus production has been found to occur when air moisture levels are below saturation point and successful healing is reduced. Air moisture levels can be maintained by sealing immediately around the graft or placing the graft in an atmosphere of

100 percent humidity. The particular treatment depends on the graft used.

Grafts like whip and chip budding can be sealed at the graft union. Using polythene or parafilm seals the cut as well as applying pressure. If rubber ties are used, then ties need to be sealed with wax. Horticulture grafting wax becomes liquid at a low temperature and so will not damage the cells of the grafted plants. Paraffin (candle) wax has also been used effectively, even though it has a higher melting point. If this type of wax is used, then it is important to dip the graft quickly in and out of the wax. Where a shield graft like T-budding is used, desiccation can be prevented by using a rubber patch to hold the bark in place over the scion bud.

For some types of graft like the side-veneer and its variations, sealing the union is not possible. Here desiccation is prevented by placing the grafted plants under a polythene tent to maintain a high humidity around the cut surfaces. This is usually done within a greenhouse.

Temperature

The optimum temperature at which a graft union forms depends on the species being grafted. In low temperatures, the graft union does not develop or it forms slowly, while in high temperatures, the scion may lose moisture and fail to grow, or produce an abundance of soft callus growth that can be easily damaged. Apples have a minimum temperature of $0°C$ and a maximum of $32°C$, while the optimum temperature for the rapid development of callus is $25–30°C$. Grapes will not form a union below $15°C$ and have a maximum of only $29°C$, while their optimum temperature is between 24 and $27°C$.

With summer budding, the higher air temperatures mean that the graft union will form rapidly. Winter grafting in the field can be used but often gives variable results due to low temperatures and inclement weather making the grafts slow to take and prone to damage. Bench grafting enables both; it provides some protection to the grafts so that the union can form slowly at low temperatures, or the optimum temperature can be applied at the point of union through a hot callus pipe. The hot callus pipe applies warm air at the graft union but keeps the scion buds cool. This encourages the union to form in only three weeks, but keeps the scion buds dormant. The use of bench grafting in the summer is becoming more popular as the scion buds are dormant at that time, but temperatures are near optimum for the union to form.

Where the graft is put under polythene to control humidity, the temperature and light levels are important. The measure for total light during a day is megajoules per square metre (MJ per sq m). In temperate zones, winter light may be as low as 2 MJ per sq m, with a maximum of about 20 MJ per sq m in midsummer. Below 1.5 MJ per sq m, there will be no net gain in carbohydrates, that is, a plant will use more energy than it will gain from photosynthesis. Above 5 MJ per sq m, light energy will be in excess from that which can be used for photosynthesis and growth, and the excess light levels will produce heat energy and raise the temperature around the grafts. In addition, under a polythene tent, as the air temperature increases, the relative humidity will drop.

It is important, therefore, to ensure that enough light reaches the graft union for photosynthesis to occur, but that the light level does not increase the temperature too much. At its extreme, high temperature will scorch the leaves on the graft and lower the humidity, causing desiccation of the scion. If the temperature is only slightly too high, the problem is slightly different in that the scion buds will start to grow before the union has formed, usually at $18–21°C$. This growth will eventually collapse because it is not connected to any roots that can support it. Shading levels similar to those recommended for rooting cuttings (under low polythene) will probably give the correct conditions.

Records

It is important to keep good records during the grafting process so that grafting can be replicated the next year if successful, or modified in a controlled manner if the results are less than satisfactory. In commercial settings, it is important to record each grafter's percentage of live grafts per hour. One grafter may do fewer grafts per hour, but, because the grafting is done better than someone grafting faster, the first grafter may produce more live grafts overall.

To determine the percentage of live grafts per hour, record both the number of grafts a person prepares per hour and, later, the number of grafts that successfully grow. Then divide the number of successful grafts by the number of grafts made and multiply by 100 to get the percentage of live grafts per hour.

One system of record keeping may include the following information, depending on type of graft:

Scion:
  Genus, species, cultivar
  Collection date, condition

Rootstock:
  Supplier
  Genus, species, cultivar
  Stem diameter, root condition
  Preparation, sap flow
  Time of year and specific date

Grafter:
  Person who prepared the graft
  Percentage success rate recorded in summer
  Number of live grafts per person

Where put after grafting
Weather conditions: daily and seasonal
Temperature: maximum and minimum
Grafting tape and source
Grafting tools used in operation
Daily humidity/rainfall
Number of plants propagated
Time of side shoot removal
Aftercare of plants post grafting
Pest and disease

4

Production of Rootstock and Scion Material

NO MATTER HOW SKILLED A GRAFTER IS, the plants produced can only be as good as the parent material used. The scion material must be true-to-type, healthy with the buds at the correct stage of development. The rootstock must also be true-to-type, healthy with cell division actively occurring. It is not only fundamental to the success of a graft but also to the subsequent quality of the plant produced.

Most propagators obtain their rootstocks from specialist suppliers. It is important to know what species and cultivar of stock is desired, the diameter of the stem, and what type of rootstock is required, whether bareroot, cell-grown, or pot-grown. In addition, the rootstock needs to be free from pests and diseases, and have a fibrous root system and a stem that is straight, free from side shoots and not excessively woody. Alternatively, rootstocks may be home produced, and so some

65

aspects of propagating rootstock material from seed, layering, and leafless winter cuttings (hardwood) will be discussed in this chapter.

Acquiring rootstocks from growers

When obtaining rootstocks from a grower, it is important to give precise specifications for the requested material. The term *pencil thick* is often used to describe the diameter of the rootstock required, but this is an imprecise term and the actual diameter of the rootstock required should be specified when obtaining rootstocks. Different countries have their own specification standards. In Europe, refer to European Technical and Quality Standards of Hardy Nurserystock, produced by the European Nurserystock Association (ENA). In the United States, the American Nursery and Landscape Association publishes the American Standard for Nursery Stock. Both documents include sections on rootstock specifications and should be utilized when ordering plant material. The standards designate diameter measurements for the girth of the rootstock as "root collar diameter," which is the diameter of the main stem measured at, or within, a specified distance from the root collar. The actual girth you specify in your order will depend on the species being grown, how and when it is to be grafted, and your location. For rootstocks to be top worked, an indication of working height is also required.

Seedling age and condition
Seed-raised rootstocks are traditionally grown outdoors in the field. They may spend one or two years in the seedbed and then can be transplanted (lined out) to grow on for up to another two years. In Europe, seedling age is indicated in the plus system for specifying this type of material. For example, 1+1 (may also be written 1/1) means the plants were grown one year in the seedbed and then transplanted and grown for an additional year. Seedlings must be transplanted after no more than two years to give the plants more space to develop and to allow the grower opportunity to prune the roots and thus encourage a fibrous root system. Transplanting a seedling before it is two years old enables the plant to establish better when it is finally potted or planted out.

In some cases, plants are undercut instead of transplanted after one year. Undercutting prunes the taproot, removing the apical dominance of this root and encouraging fibrous roots to develop. The process is particularly important with plants such as *Fagus sylvatica* (beech) that have a very dominant taproot and can suffer considerable transplant shock if not undercut. Undercutting is indicated by the use of *u* instead of + when indicating seedling age, as in 1u1.

Where plants have been propagated by layering or from cuttings, a 0 indicates this. A one-year-old layered plant is therefore denoted as 0+1.

Stem diameter
Rootstocks are ordered within a range of 2–3 mm for the stem diameter, and so come as girths of 5–7 mm or 6–8 mm, and so forth. Determining what girth size to order depends on the species, the rootstock's development during preparation for grafting, and possibly its growing conditions. Stock for immediate bareroot bench grafting needs to be up to 10–12 mm or even 18–20 mm for heavy stocks. For stocks that will be lined out for budding in the summer, a smaller rootstock is desirable, perhaps 6–8 mm in diameter, as such rootstocks will thicken up during their growth prior to budding. Rootstocks that will be grown on in pots for bench grafting in the summer or

following winter can be ordered as small seedlings of 4–6 mm, as they will increase their girth considerably after being potted up.

Growing conditions may also affect choice of rootstock diameter. Bush roses are always sold as maidens; that is, they are budded in the summer and grown for one season. If they do not make the grade at the end of the first year, they are discarded. In one climate a rootstock of 3–5 mm might be used to produce the required standards in the maidens, while in a colder climate, where the growing season is shorter, a larger rootstock of 5–8 mm is necessary.

Growing rootstocks from seed

The seeds of most temperate trees and shrubs will not germinate immediately after ripening, even if they are given optimum germination conditions. These seeds are said to be dormant, and require particular conditions to occur before they will germinate. Seed dormancy has evolved as a survival mechanism, preventing germination at an unsuitable time, such as autumn, and spreading germination over many weeks or even years. It may also be involved in spreading seeds over a wider area, away from the parent plant.

Table 1. Suggested Specifications for Rootstocks.

AGE SPECIFICATION	TYPE OF PLANT
0+1	1-year stock from stool or cuttings (normally graded 5-7mm, 6-8mm, or 8-10mm)
0+1+1	2-year stock bedded for 1 year or transplanted (normally 6-8mm, 7-10mm, 8-12mm)
1+0	1-year seedling for potting or lining, (normally graded 4-6mm, 6-8mm, with a leader if specified)
1u1	2-year seedling undercut at the end of the first year (normally 5-7 mm, 6-8 mm, or 8-10 mm, or plants with leader for lining out to grow on)
1+1	1-year seedling + 1 year transplanted (normally 5-7 mm, 6-8 mm, 8-10 mm, or 10-12 mm, or plants with leader for lining out to grow on)
1+2	1-year seedling + 1 year transplanted (normally 8-10 mm or 10-12 mm for heavy stocks; larger stems 60-150 cm in diameter for top working)
1+1P	1-year seedling, potted for 1 year, normally in P9 or 1-litre pot for grafting or lining
1+0P	1-year seedling pot grown for grafting or lining
Cell 100cc	1- to 2-year-old plants from seed with a stem 3-4 mm in diameter
Cell 200cc	1- to 2-year-old plants from 50-cc cells are transplanted to 200 cc to grow a 2- or 3-year-old plug plant, 4-5 mm girth, that can be used as a rootstock instead of a P9

Source: Adapted from Wood (1996).

Plants that produce fruits, berries, or hips contain seeds with hard seed coats. Birds or animals are attracted to eat the fruits, and in the time it takes for the seeds to pass through the gut, the animal has moved a distance from the parent tree. A hard seed coat is required to prevent the acids in the gut from damaging the embryo. Until the hard seed coat is removed, however, the embryo cannot absorb water and germinate. In nature, these methods of seed dispersal in time and space are extremely successful for species survival, but they are also very wasteful, with far less than one percent of seed ever growing to a mature plant.

Pre-treatments to overcome dormancy

Controlled pretreatments to overcome dormancy are required to obtain the maximum number of plants that meet the specifications required of rootstocks. Some species only require a cold treatment to germinate. *Abies alba, Acer platanoides, A. pseudoplatanus, Betula pendula, Corylus avellana*, and *Fagus sylvatica* are examples. Other species have seeds with hard seed coats and also have embryos requiring a cold treatment. Among these are *Carpinus betulus, Fraxinus excelsior, Rosa dumetorum* 'Laxa', *Sorbus aucuparia*, and *Tilia cordata*. It is important to note that species with both types of dormancy require the hard seed coat to be removed before the embryo can respond to the cold treatment.

Seeds with hard seed coats require a pre-treatment called scarification. This often involves a violent treatment: cutting, abrasion, or soaking in hot water or even acid. For some species, these methods can be effective. Members of the legume family such as *Laburnum anagyroides* can be scarified successfully by chipping. To do this, secure the seed (pressing on the seed with an eraser is recommended) and, using a sharp knife or scalpel, chip away a small section of seed coat.

Alternatively, seeds with hard seed coats can be treated successfully by hot water. The seeds are softened by adding them to a large volume of water three to ten times the volume of the seeds. The water should be at boiling point but removed from the heat source before the seeds are added. The seeds are left in the water for 18–24 hours until the majority of seeds have swollen through imbibition.

Acid scarification can also be used, where seeds are put in sulphuric acid which eats away the hard coat. Sulphuric acid is a very dangerous material to use, however, and requires specialist facilities like fume cupboards to carry out safely.

For most species, these treatments give varied results, leading to fewer seeds germinating than was their potential. The seeds that do germinate do so over a long period, so that by the end of the growing season, seedlings are variously sized. Differences in the effectiveness of these treatments are due partly to genetic differences between species and partly to differences between batches of seeds of the same species. The latter type of variation occurs because the hardness of the seed coat is also affected by the degree of seed maturity at collection, the air humidity during the ripening process, and the time the seed is in storage.

To overcome some of the variation occurring with traditional scarification techniques, a warm treatment was developed to mimic conditions in the wild that cause the natural biological decay of the seed coat. The seeds must be kept moist to prevent desiccation of the embryo. They must also be aerated, as there is considerable respiratory activity within the seed requiring oxygen to avoid damage and even death to the embryo. The seed can then be kept at between 20 to 25°C to overcome the hard seed coat. Warm treatment produces a gentler and more controllable removal of the hard seed coat.

Once any hard seed coat has been removed, the seeds can be stratified, that is, given a cold treatment to activate the embryo. Stratification involves changes in the biochemistry within the embryo. In plum seeds, the levels of abscisic acid (a growth inhibitor) decline during stratification, while levels of gibberellin (a growth promoter) increase to a point when the seeds can germinate. Stratification will only be effective when the embryo has absorbed enough water to activate these internal processes.

Again, a number of techniques are used to stratify seeds, from autumn sowing to the use of stratification pits. These techniques can be used effectively but still give varied results that can reduce the germination percentage of a seed lot, and/or spread the germination over a long period. To provide more control over stratification, a similar approach to the warm treatment is used, except that the seed is maintained between 1 and 5°C for the required time. Because of the variability in treatment times, it is important to keep records so that treatment times can be refined each year.

Even when using warm and cold pre-treatments, seed germination will be spread over several weeks, resulting in variously sized seedlings by the end of the season. Dormancy is not a light switch where all the seed is dormant one moment and then able to germinate the next. The treatment times recommended in Table 2 are frequently given over a spread of weeks. The actual time needed depends on the source of the seed batch, weather conditions during ripening, and storage conditions. If stratification is carried out for the shortest recommended time, many seeds may still be dormant when sown, delaying their germination. This will lead to germination over a number of weeks and give seedlings of variable sizes. If, however, stratification is carried out for the maximum time recommended, then a significant proportion of seeds may have germinated during stratification. This will cause them to be etiolated and the emerging radicle may be damaged when the seeds are sown.

Cell-grown vs. field-grown

In recent years, there has been a move from growing trees and shrubs in seedbeds in the field to growing them under protection in cell trays. Once again, terminology can be confusing. *Cell-grown* refers to plants grown in trays divided into sections. In bedding plant production, these containers are usually referred to as modular trays, but for tree seedlings, the term used is *cells*. Cell-grown plants may also be referred to as plugs or plug plants.

Tray design for trees is different than for bedding plants: the cells are deeper and tend to be ridged to force the roots to grow down and not to spiral round the pot. Tree cells are also open at the base and raised off the ground. One way to elevate tree cells is to put the trays on upturned pots, but they can also be grown on a low table that has a mesh top. In either case, the aim is to air prune the roots. The taproot emerges through the bottom of the cell and the tip is killed by coming directly into contact with the air around the bottom of the tray. Pruning the taproot in this way promotes secondary rooting and produces a well-formed fibrous root system.

The advantage of cell-grown trees for forestry plants is that cells reduce the time seedlings are in production as well as the risk of trying to predict demand several years ahead. By using the correct modules and root pruning techniques, planting from cells is easier and gives better establishment of young trees. Some nurseries also use this production method to grow rootstocks of the correct height, girth, and root system required by grafters.

A disadvantage of cell-grown trees over field-grown seedlings is the cost of production. Investment in greenhouses, irrigation systems, cell trays, and precision sowing is required. Since the germination percentage can be well below 100 percent, sowing a single seed into a cell often leads to many gaps in the crop. Sowing multiple seeds into a cell ensures that at least one of them will grow, but wastes seed and requires the seedlings

to be thinned to one per cell. Alternatively, seeds can be sown in small cells initially and then transplanted to the final, larger cell. Again, however, this method adds a handling stage and a cost to growing the crop. Two approaches are being taken to improve the evenness and percentage of germination: the use of controlled moisture stratification and techniques to separate live from dead seed.

Table 2. Standard treatment times for warm and cold stratification of selected species used as rootstocks.

DECIDUOUS SPECIES	WARM STRATIFICATION AT 10–15°C (NUMBER OF WEEKS)	COLD STRATIFICATION AT 1–5°C (NUMBER OF WEEKS)
Acer palmatum	4	4–12
Acer platanoides	0	8–16
Acer pseudoplatanus	0	6–12
Berberis thunbergii	0	6–13
Carpinus betulus	4	12–14
Crataegus monogyna	4–8	12–16
Fagus sylvatica	0	8–12
Fraxinus excelsior	8–12	8–12
Prunus avium	2	18
Quercus rubra	0	4–8
Rosa canina 'Inermis'	8	8–12
Rosa multiflora	0	12–18
Sorbus aucuparia	2	14–16
Syringa vulgaris	0	4–12
Tilia cordata	4–20	16–20
CONIFER SPECIES		
Abies alba	0	3–6
Abies nordmanniana	0	3–6
Pseudotsuga menziesii	0	3–6

Note: The time for each treatment will depend on a number of factors including provenance, conditions during ripening, and storage conditions and time.

Controlled moisture stratification

A seed embryo has to imbibe moisture before it can respond to cold stratification. Seeds are therefore soaked in water prior to being stratified. This means, however, that once the required cold period has been reached, the seed will germinate. With the seed treatment described above, once 10 percent of the seed has begun to germinate, then the whole batch must be sown. Stopping the stratification process, however, may result in a significant amount of seed remaining dormant that will either be slow to germinate or fail to germinate once sown.

Studies in stratification have shown that there is a point where moisture content in the embryo will enable the stratification process to proceed but is below the required moisture content for germination. This is called controlled moisture

stratification and has led to the identification of the critical moisture percentage for a range of species. Niels Dictus, who offers a stratification service to growers in the Netherlands, has identified moisture content percentages for a range of species. He found that seed of *Acer platanoides,* cold stratified in the normal way for eight weeks resulted in around 30 percent germination. Longer stratification can achieve about 40 percent germination but risks the loss of seeds germinating earlier during stratification. For controlled moisture stratification of this maple species, a moisture content of 36 percent needs to be maintained; 34 percent is too low and will lead to the seeds drying out and not responding to the cold treatment, while 38 percent is too high and will trigger premature germination.

To achieve an optimal germination rate for *Acer*

Tree seedlings being grown in cells under protection. The more intensive production reduces the production time from sowing to selling the rootstocks. The cells are ribbed to prevent root spiraling. This makes a fibrous straight root system for better establishment once the seedling is planted out.

platanoides, the seed is split into small batches, weighed, and placed in bags (small batches are easiest to check during stratification). A sample is taken from each batch, weighed, and placed in an oven at 105°C for 17 hours. This temperature and length of time in the oven allows the seeds to dry without affecting other constituents like oils. The samples are weighed again at the end of the drying period, and the difference between the original weight and the dried weight gives the moisture content as a percentage of the weight of the seed sample. This percentage is used to calculate the amount of water required to bring the moisture percentage up to the required level of 36 percent for the seed batch.

After the required quantity of water is added to the seed batches, the seeds are mixed and then stratified naked at 4°C. Once a week, thereafter, the seeds should be remixed and their moisture content checked. If the seeds are sown after 15 weeks of this treatment, a germination rate of about 34 percent is achieved, which is in line with traditional stratification. However, since the seeds that have broken dormancy will not germinate, stratification

Carrying out warm and cold treatments

STEP 1 To determine when treatment should begin, start with the date seeds are to be sown and calculate back the number of weeks required to complete the treatment. Plan to allow for an average treatment time, not the shortest nor the longest for the species. Keep records to refine treatments each year. Sow seed outdoors when the soil temperature is about 10°C; in some areas this temperature may be reached in early spring, elsewhere in late spring. Avoid sowing too late, as seedbeds are likely to require irrigation and the season may be too short for the seeds to make the required growth in a season.

STEP 2 Divide the seed into manageable batches, place it into cloth bags, and soak each bag in several times the volume of water to the bag at 3–5°C. A refrigerator could be used to maintain the proper temperature.

The aim here is to wash and remove any chemical inhibitors as well as to enable the seed to absorb water if it does not have a hard seed coat.

Remove the bag from the water regularly and allow it to drain. Replace the water if it is discoloured before re-submerging the seeds. Drain off the water after 48 hours.

The easiest way to control the pretreatment of seed is to mix it with a medium that retains moisture (thus preventing the seed from drying out) and provides air (thus allowing the seed to respire). To prepare such a mixture, combine an organic, moisture-retentive material such as peat or leaf mould with an equal volume of an inorganic material such as grit or perlite, and wet the mixture until drops of water can just be squeezed from it through your fingers. Mix the seed thoroughly with the medium.

can be continued up to 19 weeks at which point germination results are about 78 percent.

Fagus sylvatica is stratified at 30 percent moisture content and shows a germination percentage of 54 percent after the traditional ten-week period. If controlled moisture stratification is continued for another six weeks, the germination rate increases to 82 percent. In addition, the germination will occur over a short period as dormancy has been fully removed. Recommended moisture contents and cold stratification times are given in Table 3.

Separating viable seed from dead seed

Controlled moisture stratification has significant benefits, especially for cell-raised trees, but direct sowing will still lead to around 20 percent of empty cells. To overcome this problem, growers are increasingly using techniques to separate live from dead seeds. There are a number of ways to do this.

In one technique, gravity tables are angled and seed is put on them at the top. As the table vibrates, the seed moves down the table with the heaviest seed reaching the bottom and being collected first and the lightest seed being collected

Alternatively, the seeds can be surface dried on a fine mesh to allow air movement around the whole seed before the seed is put into a polythene bag to be treated "naked." In either case, the bag is tied loosely to enable gas exchange while preventing the seeds from drying out.

Treating the seed "naked" prevents the embryos of non-dormant seeds from absorbing further moisture in the absence of the moisture-retaining medium. As a result, the seeds that break dormancy first do not immediately germinate and so the pre-treatment can be carried out for longer, giving better and more even seed germination once sown. The "naked" stratification of seed, however, requires more skill to carry out successfully than using a moist medium.

STEP 3 If seeds are to receive a warm treatment, keep them at a constant 10 to 15°C for the required number of weeks.

STEP 4 Once the warm treatment is complete, or if seeds only require cold treatment, place seeds into a refrigerator or cold store at 1–5°C.

STEP 5 Open the bags every week, mix the seeds, and spray them with water if there are any signs of drying out. Remove any seeds that are decaying, as these will cause fungal infections to spread.

STEP 6 When around 10 percent of the seeds show signs of germination, remove all the seeds from the cold store, separate them from the medium (if possible), and spread them out on trays. A cool, well-ventilated shed with no direct sunlight or artificial heat is the ideal place to surface dry seed, making sowing evenly much easier. If a large proportion of the seeds have started to germinate, however, they cannot be surface dried and need to be sown immediately.

last. In general, the lightest seeds are the least viable and are discarded.

Flotation can also be used to separate many seeds. In these cases, seeds can be added to water and because the filled, viable seeds have a greater specific gravity than water, they sink. Empty dead seeds have a lower specific gravity than water and float. However dried, viable seeds often float until they have imbibed water, and so time must be allowed for this to occur before separating the seeds. This technique is known as the absorption method for seed separation.

Dead empty seed can sometimes absorb water and also sink, as will seeds that have developed

Table 3. Controlled moisture cold stratification of selected species used as rootstocks.

DECIDUOUS SPECIES	COLD STRATIFICATION MOISTURE CONTENT* (% FRESH WEIGHT)	TIME (NUMBER OF WEEKS)
Acer palmatum	35–37	12
Acer platanoides	35–37	19
Acer pseudoplatanus	46–48	15
Berberis thunbergii	38–40	14**
Carpinus betulus	28	16
Crataegus monogyna	28–30 (Denmark) 22 (Italy)	24
Fagus sylvatica	30	20
Fraxinus excelsior	42–43	14**
Prunus avium	28	18
Quercus rubra	36–37	10**
Rosa canina 'Inermis'	25	14**
Rosa multiflora	36–37	20**
Sorbus aucuparia	43–45	18**
Syringa vulgaris	45	14**
Tilia cordata	42	24
CONIFER SPECIES		
Abies alba	32–34	8**
Abies nordmanniana	32–34	8**
Pseudotsuga menziesii	34–36	8**

Note: Standard cold stratification temperature is 1–5°C and follows warm stratification where required.
*Provenance of seed can affect optimum moisture levels. See *Crataegus monogyna*.
**Estimated time from traditional stratification guidelines.

embryos but have been damaged in some way and are no longer viable. To overcome this, the absorption method has been refined. In this case, the seeds are re-dried from a few minutes to several hours. During this drying process, the empty and damaged seeds will lose water more rapidly than healthy seeds. The flotation test is repeated, and the healthy seeds sink while immature or damaged seeds once again float.

Several other techniques may be of value in separating viable from non-viable seeds in the future. Three have developed from the absorption method: incubation-drying-separation (IDS), pressure-vacuum (PREVAC), and density method. IDS and PREVAC have been used with pine seeds but may have wider applications. Both techniques based on the principle that live seeds absorb water and dry at different rates than dead or damaged seeds.

IDS is mainly used with stored seeds that may have lost viability prior to sowing. In this method, the seeds are soaked to imbibe water and then put into an optimum germination environment—light, temperature, and high humidity. These

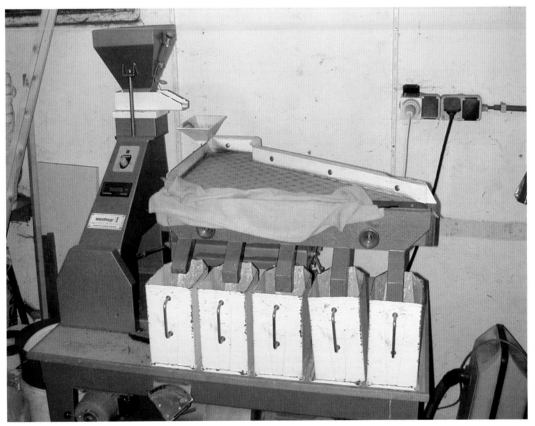

Gravity table that separates seed by weight. Where an embryo has developed and filled the seed, the seed will weigh more and be more likely to be viable compared with lighter seeds where the embryo has not fully developed.

conditions are maintained for two or three days, after which time the seeds are partially dried. The flotation test is then carried out when the dead seeds, which tend to lose water faster during drying, will float, while live seeds will sink. The problem with this method is that germination may begin during the two or three days in the optimum germination conditions. This is why it is used after seed storage prior to sowing. Even so, germinating seeds may be damaged if dried back too far.

The PREVAC method is used to separate seeds with mechanical damage from sound seeds. Dry seeds are placed in water in a container and a partial vacuum created for 1–20 minutes. On release of the vacuum, damaged seeds absorb water more quickly than undamaged ones. This is because

there will be cracks or even parts of the seed coat missing. When the floatation test is carried out, the less viable, damaged seeds sink and are subsequently removed.

The density method is another adaptation of the flotation technique that uses liquids of different densities to give separation that is more precise. The specific gravity is adjusted between that of the heavy and lighter materials to be separated using different types of alcohol or ethers, or by adding salts to water. Soaking in these materials runs the risk of damaging the embryos, so they are not widely used.

Other novel separation methods are being investigated like chlorophyll fluorescence (CF) and near-infrared technology (NIT) that may

After controlled pre-treatments, seeds are coated with a vegetable dye so that any gaps during sowing can be seen and gapped up.

have potential as non-destructive methods of assessing seed quality. In some agricultural crops, like brassicas, live seeds can be separated from dead by CF due to changes in the chlorophyll content of the seed coat. CF may have potential for use with tree species also. NIT is used to scan individual kernels of cereals with light in the near-infrared range. This method detects characteristics such as crude protein, starch, and moisture that are important to the users of cereals in food products. NIT may be useful to assess seed quality for germination potential.

The developments in pre-treatments to overcome dormancy and in techniques to separate viable seed from dead seed means that rootstock quality from seed-raised material, in particular from cell-grown plants, should continue to improve and meet the requirements of the grafter.

Obtaining rootstocks from layering

The ability of a stem to develop roots depends on two factors: *totipotency*, the fact that all living cells contain the genetic information to form a new plant and its functions, and *dedifferentiation*, the ability of cells to lose previously developed functions and form a new meristematic growing point. Adventitious roots may develop on a stem from preformed root primordia (an organ or tissue in its earliest recognizable stage of development) present in the stems, or from wound roots that may develop if a shoot is severed from the parent plant. Both types of roots will only develop if given the correct environmental conditions.

Layering involves the rooting of stems while they are still attached to the parent plant. There are several different methods of layering plants depending on how they grow. For fruit rootstocks, mound layering is commonly used. It is carried out from stock plants that have initially been established by stooling or by a form

of trench layer. Mound layering depends on the ability of fruit rootstocks such as apple, pear, and cherry to produce new shoots after the stem of the parent plant has been hard pruned to the ground. Given the correct environment of warm, moist, dark conditions, these shoots will contain preformed root primordia that can develop into a fully formed root system. At the end of the season, the rooted propagules can be removed from the parent plant and grown on for use as rootstocks.

Establishing the layer bed can be achieved by one of two methods. Stooling involves cutting back the parent plant to ground level once it has established for a season. This commonly used method produces individual clumps of shoots that can be mounded up. An alternative method is to plant the parent plant at an angle of 45 degrees and, after one year, peg the main stem down to the ground. The stem then produces a more continuous row of new shoots that can be mound layered. Frank P. Matthews' nursery has found that stooling produces a much tougher clump of propagules to cut through; it cuts stools with a disc blade.

The type of rooting, produced by mound layering, will also occur in apples when grafted material is planted in the orchard. If grafted plants are planted deeply, then adventitious roots will form up the covered stem and this can improve the anchorage of newly planted trees. If the scion is covered at planting, it can also form roots, but this should be avoided as the benefits of vigour control from the rootstock will then be lost. For this reason, the scion of fruit is grafted higher up the rootstock than the scion of ornamental trees. The graft is purely a method of propagation with ornamental trees, and it is not a problem if the scion develops its own roots. A new apple rootstock, MM116, has been released to replace MM106, as it is more resistant to *Phytophthora*. This resistance seems to be due partly to the stem not producing adventitious roots after it is planted. Unfortunately,

MM116 is proving more difficult to propagate because it does not produce roots easily.

Mound layering

Although layering is used to produce large numbers of rootstocks, it can also be carried out on a small scale for relatively little investment. Layering produces a large plant that may be used for grafting or budding the season after lifting. Since the parent plants will produce layers for many years, it is vital that high-quality, true-to-type material is planted in the first place.

Selecting a suitable site for a layer bed is important since it is a long-term investment. The site also needs to be prepared correctly prior to planting to avoid problems in the future. Ideally it should be sheltered and facing the sun, with a slightly sloping terrain that it will warm up early in the spring and provide plants with as long a season as possible for root development. The soil should be well drained with a sandy loam texture, free of stones, so that it will be easily mounded up around the developing stems. Most importantly, the soil should be free from pests, like potato cyst nematode, and diseases like *Verticillium* wilt. It should also be free of weeds, particularly perennial weeds such as creeping buttercup that could become major problems over the long term of the layer bed.

An ideal site may be unavailable, but pre-planting preparations will help improve an area. For example, drainage can be improved if required; organic matter can be added to a soil to improve its texture; and perennial weeds must be removed before planting. It is also important to ensure that the soil is kept moist during the growing season and so irrigation should be available. Soil pH should be 5.5–6.5 and nutrition levels in the soil tested.

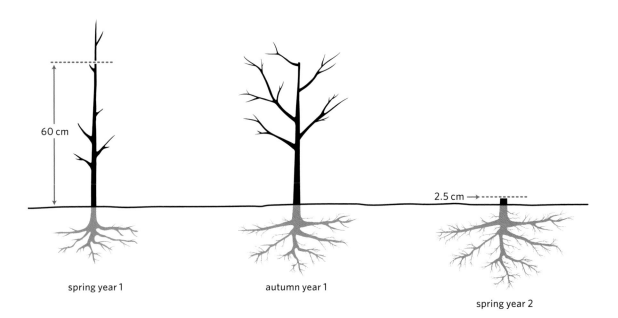

Figure 4-1. Stock plant establishment (left to right): Stock is planted in spring and headed back to 60 cm. By autumn, the stock plant has established. In the spring of year 2, the stock is headed back to 2.5 cm above soil level.

Stool bed technique

The selected rootstock should be obtained from a reputable supplier and preferably should be certified stock. It is planted in spring on a north–south access to gain the maximum summer sun and high soil temperatures, which will give the longest growing season. Plants should be 25–40 cm apart within rows, depending on vigour, and 1.1 m apart between rows. Spacing the plants as closely together as possible will maximize the productivity of the beds. After planting, the stock is pruned to 60 cm and allowed to grow for a season before stooling. This will establish a strong root system by autumn.

Once leaf fall occurs in autumn, the stock is cut back to 2.5 cm above the ground (Figure 4-1). Because a clean cut is very important at this stage, often a knife is used rather than secateurs. If damage occurs at the cut, then rot can occur and affect the productivity of the stool. It is important to prune the plants in autumn, so that the latent buds around the base of the stem will be left exposed. These receive a cold chill (vernalization) during the winter and grow evenly in the spring. In areas of particularly low winter temperatures, it may be necessary to protect the stools by covering with soil or straw to insulate the buds.

Once the buds start to grow in spring, the soil is mounded up around the stock (Figure 4-2). This can be done twice, but three times appears to give the optimum results as far as rooting stem development is concerned. It is important that the mounding process cover no more than half the length of the stem at any time. This is to prevent too great a reduction in photosynthetic activity and growth.

At Frank P. Matthews' nursery, white pine sawdust is used for the first cover in early summer

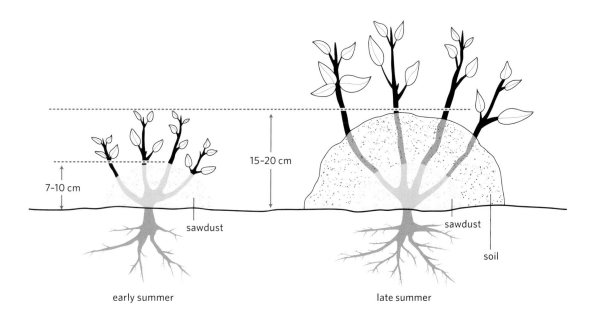

Figure 4-2. Production of layered propagules (left to right): Sawdust is used for the first mound around the stool bed. Further mounding is with soil to a height of 15–20 cm.

In Oregon, sawdust is in abundant supply because of the local forestry industry and so is used for all the mounding. It is a labour-intensive activity with the sawdust being pushed around the stems, ensuring that they remain straight and separate to allow even root development.

Layer beds in the summer after the final mounding.

when the new growth is about 15 cm long. The sawdust has been found to help produce a good root system and applying it early partially etiolates the stem, prevents lignification, and helps rooting. In Oregon, sawdust is used at each covering but this is mainly because the forestry industry in the state ensures that there is an abundance of this material available. In most areas, the soil at the side of the beds is used to mound up following the first covering. The shoots are then covered two more times so that by the end of the process, 15–20 cm of the stem base is covered. The covering is carried out three times so that the stem is covered as quickly as possible as it grows, giving the roots the maximum time possible to develop before the winter.

For roots to develop, the sawdust or soil must be closely packed around the stem to give a dark, moist environment with no air gaps that will inhibit root growth. The soil must be a fine tilth with no large lumps where air gaps can occur. In addition, the stems must be kept straight and evenly spaced around the stool to ensure that, when the propagules are lifted in the autumn, as many as possible meet the requirements for straight stems, stem diameter around 3–12 mm, and a good fibrous root system. To achieve this, the medium is put over the stool beds with shovels or a draw hoe, and then worked down between the stems by hand to separate the stems and keep them straight. It is a labour-intensive operation.

Husbandry during the growing season is to irrigate the beds if required. The soil needs to be kept moist. Weeds should be kept under control following the final earthing up. Before this, the process of earthing up will also control any annual weeds that are germinating. It may also be beneficial to top dress with fertilizer, especially nitrogen and potassium.

Once the leaves have dropped in autumn, the rooted propagules are ready to be removed from the stool bed. The soil covering the roots has to be carefully removed. Straw is often laid between the rows to keep things cleaner and to reduce damage to the soil structure if it has been wet. The straw also provides organic matter to be incorporated later into the soil. All the shoots that have developed during the growing season are then removed by cutting as close to the stool as possible (Figure 4-3). This even includes shoots that have not rooted, so that the stool only has buds present going into the next winter. Care must be taken not to damage the roots, which should be placed in bags as soon as removed to prevent them drying out.

Any shoots that have not rooted are discarded. The rooted propagules are graded, typically to a length of 60 cm and diameters just above the roots of 3–5 mm, 5–7 mm, 7–9 mm, and 9–12 mm. The thinner stems will probably be lined out and grown for a further season to produce a larger diameter stem. Any side shoots that have grown from the main stems are removed with secateurs.

The stool bed should once again be cut back to just above ground level and the buds allowed to vernalize over the winter. Organic materials like pulverized bark, spent hops, or compost should be applied over winter. Farmyard manures can be used also but care must be taken that they will not introduce weed seeds into the stool bed. Removing rooted propagules from the bed also removes a significant amount of nutrients from the soil each season. The application of organic materials will help replace some nutrients, but in addition, fertilizers should be applied in early spring based on a soil analysis to ensure the vigorous growth of the new shoots the following year. In this way, a stool bed should remain productive for at least 10 years.

Trench layering technique

In this technique, the stock is planted at an angle of 45 degrees or greater and established for a year (Figure 4-4). In the second winter, the stems are pegged and tied down to ground level. New shoots will grow vertically from along the horizontal parent plant so that it is a more continuous row of stems compared to the stool method. The parent plant is also planted in a shallow trench

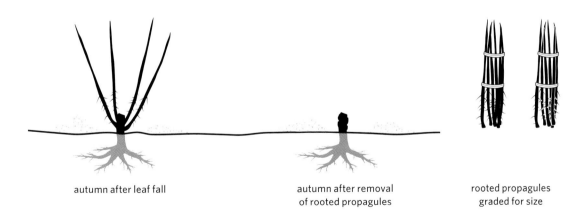

autumn after leaf fall

autumn after removal
of rooted propagules

rooted propagules
graded for size

Figure 4-3. Harvesting rooted propagules (left to right): Once leaves have dropped, the soil is carefully removed from around stool. All shoots are removed from the parent plant, taking care not to damage the roots. The layered propagules are graded to shoots greater than 60 cm and girth 5–7 mm, 7–9 mm, and 9–11 mm. Smaller shoots are discarded but must be cut from the stool.

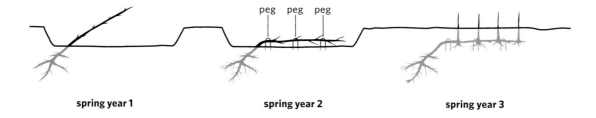

spring year 1   spring year 2   spring year 3

Figure 4-4. Establishment of layer bed (left to right): Parent plant is planted at an angle of 45 degrees in a shallow trench. In year 2, it is pegged down completely to the ground. In year 3, mound layering can commence as above.

New shoots growing up from a young layer bed. The new growth is spread out more than with a stool bed.

so that, once established, the stems are cut back to ground level each year, which is just above the original parent stem. The benefit of this method is to make the mechanical cutting of the rooted propagules easier than if a stool system is used.

Mound layering is therefore a reliable method of producing rootstocks each year. Although Matthews' nursery has managed to mechanize the process, mound layering is usually a labour-intensive activity that requires skill to ensure it is carried out correctly and the final propagules are removed with care. For this reason, growers have been keen to develop methods of rooting leafless winter (hardwood) cuttings to reduce the costs of having stool beds.

Propagating rootstocks from hardwood cuttings

An alternative method of propagation for rootstocks of tree fruits (and a few ornamental trees) is leafless winter (hardwood) cuttings. A cutting is where a stem develops its own roots after being detached from the parent plant.

Hardwood cutting is yet another horticulture term with multiple meanings. Most commonly, it refers to stem cuttings of deciduous woody plants taken between leaf fall in autumn and bud break in spring, but it can refer to stem cuttings of evergreen woody plants taken between autumn and spring. To avoid confusion in this book, *leafless winter cutting* refers to deciduous hardwood cuttings.

Given the correct environment, wound-induced roots develop at the cut base of stem cuttings. The first step in the process is de-differentiation of previously differentiated cells to become meristematic cells. Next, root initials form near the vascular tissue from the meristematic cells. Last, the root initials form into root primordia and these emerge to form the new roots as well

as an internal vascular connection with the stem. Wounding down the stem of a cutting can also promote wound-induced roots from the outward-pointing cambial salient in the callus.

Preformed root primordia will also be involved in the rooting of cuttings. In apple rootstocks, root primordia occur at the cambial ring in the branch and leaf gap, that is, at the node of the stem. In willow rootstocks, the preformed root initials develop over a wider area of the stem at the nodes and also internodal points. The presence or absence of preformed root primordia does not actually indicate how easy a species is to root from leafless winter cuttings. Other factors may affect the ability to root. For example, in apples and pears the continuity of sclerenchyma cells is related to the ease of rooting. Sclerenchyma cells are dead cells whose walls are thickened with cellulose and lignin. When these are fibres, they can form a ring of varying continuity around the stem. The greater the continuity of the fibre ring, the poorer the ability of cuttings to root. For example, M25 has 71 percent continuity and is only a fair rooting stock while M5 is an excellent rooting clone and only has 43 percent continuity.

It is also important to remember that the cuttings will always form roots at the end that was closest to the roots while the stem was on the parent plant. This principle is known as polarity and holds true no matter how big or small a cutting is taken. By convention, a straight cut is made at the base of the cutting, usually just beneath a node, while a sloping cut is made at the top. In this way, there should be no confusion as to how to insert the cuttings.

In willows, promoting rooting in leafless winter cuttings is straightforward and can be achieved outdoors in the soil to provide a dark, moist condition. In other species, rooting is more difficult and requires treatments to the stem, like wounding, and the application of rooting hormones, as well as creating a rooting environment that may involve the use of higher temperatures.

An example of a species that easily roots from leafless winter cuttings is *Salix viminalis*, used as the rootstock for high working of Kilmarnock willow. Field-grown stock plants are cut back hard in spring to produce vigorous new growth. Two approaches can then be used. Cuttings can be taken in the autumn and lined out in the field to root and grow to the correct height and diameter for grafting as rooted cuttings the following year. Alternatively, stems of the correct length and diameter can be produced and used for grafting prior to root development. The correct conditions can then be provided to both root and form the graft union at the same time. In both cases, the willow will develop roots readily outdoors in a sheltered site with free-draining, weed-free soil of a fine tilth. Inserting the cuttings through black polythene laid over the soil can also help rooting.

The black absorbs heat from the sun, warming the soil and stimulating earlier rooting, and the sheet also acts as a mulch, retaining moisture in the soil and inhibiting weed growth.

Cutting grafts have been tried for other species. At Wageningen University in the Netherlands, a technique called stenting was developed for grafting and rooting roses to be used for the cut flower industry. Often cutting grafts are possible but are less viable commercially than traditional grafts. Many rhododendrons could be grafted using cutting grafts but have a low success rate of around 50 percent at best. Growers have not taken up this technique apart from in a few examples like willow. Fruit rootstocks like Quince A, Myrobalan B, Saint Julien A, M26, MM111, and MM106 are more difficult to root, but with the correct pre-treatments and rooting environment, propagation from cuttings can be successful and is less labour-intensive than layering.

Cutting grafts of rhododendron using the easily rooted *R. ponticum* as the rootstock.

Stock plant

As with the stool beds, stock for leafless winter cuttings can be grown as individual plants, or hedges can be developed for larger production with plants 30 cm by 2 m apart (Figure 4-5). The parent plant is planted and pruned to 60 cm (as for the stool bed). In the next winter, all the side shoots are pruned back to two to four buds, and this is repeated again in the second winter. Some cuttings may be taken at this time but it is really after the third season that the framework has developed and the correct cuttings are being produced. Shoots that are over 1 m long are suitable for cuttings. Once rooted, this size of cutting will be suitable for budding the following summer.

The cutting is taken at the swollen shoot base, as close to the previous year's wood as possible, as at the base there is a rosette of buds with their associated nodal sites and the presence of preformed root initials. The shoots are then shortened from the top to give 60- to 75-cm-long cuttings. Shoots less than 1 m long should be pruned back to 1–2 buds, and large feathered shoots that are also not suitable as cuttings should be pruned to 2–4 buds. In this way, the stock plant will continue to produce cutting material for many years.

It is best to collect the cuttings in late winter, although care needs to be taken to collect them before they begin to leaf out. Alternatively, cuttings can be taken in autumn, removing any leaves present. Although cuttings taken at these times will not root as well as those taken in spring, the timing does avoid the danger of taking the cuttings too late. Between these two periods, the rooting ability of the cuttings declines significantly.

Although fruit rootstocks do not root as easily as the likes of willow, many can be rooted straight into the field. The cherry rootstock Colt has

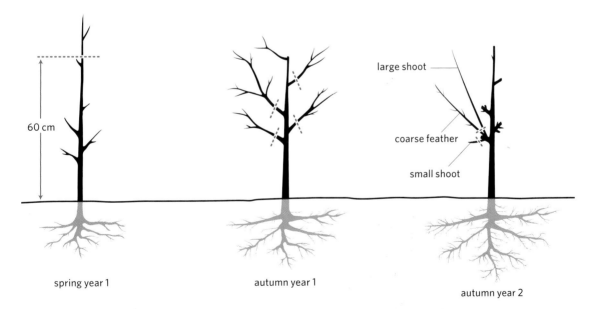

Figure 4-5. Hedge establishment for leafless winter cuttings (left to right): In spring of year 1, the stock is planted and cut back to 60 cm. In autumn of year 1, the new growth is pruned back to two to four buds and repeated in the second winter. Once plants are established, remove all shoots each year, including coarse feathers pruned to two to four buds, small shoots pruned to one to two buds, and large shoots greater than 1 m, to use as cuttings, cut as close to the previous wood as possible to retain the swollen shoot base.

Hedge of Saint Julien A, a plum rootstock, producing material for leafless winter cuttings.

Adventitious roots developing on a cutting hedge of Colt, a cherry rootstock bred to be easier to root from cuttings as well as having a dwarfing effect on the scion.

visible, pre-formed roots at the base of the stems on the stock plant. Once the stems are cut below the root initials and inserted into the field soil, rooting should be successful. Other rootstocks like Saint Julien A are less successful directly stuck into the field, but because direct rooting is a low cost method of propagation, losses are acceptable.

Robert Garner, the author of *The Grafter's Handbook* (2013), was for 30 years chief propagator at East Malling Research. As well as his grafting textbook, he is also known for the development of the Garner bin (also referred to as the Malling cutting bin) for rooting leafless winter cuttings. It is well known that the application of bottom heat for difficult-to-root cuttings is important to promote rooting. Garner developed a system that would stimulate root initiation while keeping the buds dormant prior to planting out.

Electric soil warming cables are used at a temperature of 21–22°C and these are covered by moist, medium-grade chipped pine bark. This should be constructed in an unheated shed not exposed to direct sunlight so that the air temperature is kept cool to prevent the growth of the buds (see Figure 4-6). The cuttings are treated with rooting hormone, bundled and inserted into the pine bark to a depth of 25 cm, and kept moist. The bark ensures that the cuttings do not lose moisture, but there is no water lying around the bundles to cause rot. The cuttings remain in the bark for three to four weeks, with the temperature being lowered for a few days before the end of the treatment to harden off the new roots.

It is important to note that cuttings should not remain on the base heat for any longer than four weeks and their eventual establishment may

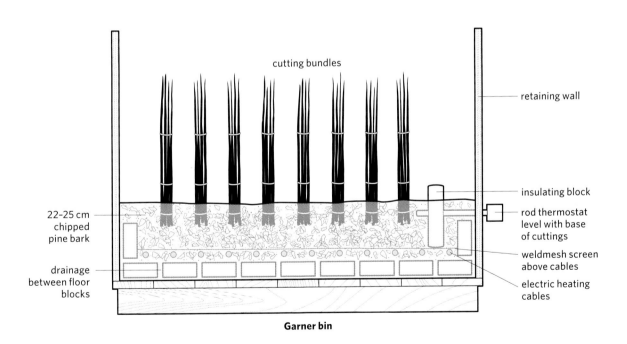

Garner bin

Figure 4-6. Construction of the Garner bin. The bin can be constructed on a pallet so it can be easily moved. It should be placed in an unheated shed in indirect sunlight when being used, or a cold store.

be better if they are given base heat for no more than two weeks. The reason for this is that the cuttings are dependent on energy stored in the stems to root and establish once planted out. The longer they are in warm conditions, the more energy they use, thus reducing their ability to grow once planted. The cuttings should have formed roots by this stage, although cuttings that have only formed callus can still be planted out—though in poor years these may not establish.

Either the cuttings can be planted out at this point or they can be kept in a cold store until favourable conditions for planting. In either case, the roots that have emerged can be trimmed back to the stem to make planting or storage easier. If the cuttings are to be stored, then they should be loosely tied in polythene bags and kept as near to 0°C as possible.

In recent years, Frank P. Matthews' nursery has moved away from using base heat for leafless winter cuttings. Instead, it prepares cuttings in early winter, wraps them tightly in thick black polythene bags, and stores them over winter in an unheated shed, at ambient temperature with no direct sunlight. By spring, roots have begun to emerge and the cuttings are ready for planting.

Producing scions

Like rootstocks, scion material must be true to type and free from pests and diseases. It may be possible to obtain graft wood from suppliers, especially of fruit, but most growers produce their own scion material by normal pruning practices. For example, in apples and other fruit trees, the vigorous, upright vegetative shoots that are suitable for budding and grafting are removed from the tree when it is pruned for fruit production. It is acceptable to collect these shoots in the summer to use as bud-sticks or to collect them for winter grafting as part of the normal pruning process.

With ornamental trees, however, more severe pruning is required to produce suitable scion growth than would normally be carried out in the

Leafless winter cuttings taken in early winter, wrapped in polythene, and kept in a cool shed that will not promote bud break. Note callused and swollen base showing signs of root emergence.

Stock plants of birch scion material heavily pruned prior to spring growth to produce new vigorous stems suitable for grafting. Some shoots are left on the tree to act as a sapdrawer early in the year.

Vigorous stems have developed excellent bud-wood material by late summer for chip budding.

Severely pruned stock plant of ornamental cherry in early spring.

garden. Severe pruning ensures that strong vegetative shoots develop and that they will be of the correct quality for subsequent budding or grafting. It may be beneficial to leave some light shoots on a tree after pruning to act as a sapdrawer in the spring. Conifers are not pruned, apart from removing the required scion-wood.

With shrubs, hard pruning to produce the correct graft wood may not affect the aesthetic appeal of the plant in the garden. The new vigorous growth will retain the correct shape and the shrub should flower as long as the pruning is carried out at a suitable time.

It is important that scion material is collected when the buds are at the correct stage of development. It is also important to collect material with the correct growth habit. *Topophysis* is the persistence of non-genetic effects after grafting due to age, position, or growth habit on the parent plant. This means that if a lateral branch of a woody plant is grafted, the resultant scion growth is often the same non-vertical growth it showed while it was still attached to the parent tree. Tying the leader to a cane will not overcome this deficiency; as soon as the branch gets above the height of the cane, it will revert to its horizontal growth.

I once collected graft wood of a fastigiate *Ginkgo biloba*. Unfortunately, all the suitable material for grafting was growing more laterally than vertically and, once grafted, produced a tree of normal, rather than upright growth habit. The same principle is true of weeping cultivars of trees, in which case the more upright-growing shoots should be avoided.

Hamamelis (witch hazel) and many other shrubs can be pruned and used for scion-wood, but will still flower and be an attractive garden plant.

The collection of *Fagus sylvatica* 'Atropurpurea Pendula' scion-wood must ensure that the shoots are pendulous and are not growing horizontally or upwards, as in the shoots to the left.

Graft wood of apples and pears being kept fresh and dormant in a cold store from midwinter until grafted in early or even mid-spring.

For summer budding, scion material should be collected as near to the time of budding as possible. Leaves are removed and the bud-sticks wrapped in moist hessian (burlap) or similar material to prevent them from losing moisture. The scions should also be refrigerated or kept in a cool box prior to taking out into the field to bud.

For winter grafting of deciduous plants, scions are best collected in midwinter prior to grafting. Desiccation is not such a problem once the leaves have dropped and temperatures are cooler, so the time between collection and grafting is not so critical. It is more important to ensure that the buds are fully dormant at the time of grafting. Collecting scions in midwinter is recommended. The sticks can then be stored in a refrigerator at below 5°C to ensure the buds remain dormant. Winter grafting of evergreen material, like conifers, is best collected fresh just prior to grafting, although it can be stored for a few weeks if required.

It is often recommended that, when cutting the scion, a nurse bud should be retained behind the whip cut. The reason for this is that the bud is an area of meristematic cells that readily dedifferentiates and forms the callus and then the new cambium connection. This is not an absolute requirement, however, and while some growers prefer to have a bud behind the scion cut, others do not find it affects the success of the graft significantly. They simply cut their scions to length and cut at a convenient piece of the stem. Where a species produces long internodes it may not be possible to use a nurse bud and the graft will still successfully unite, for example, *Acer pseudoplatanus* 'Brilliantissimum'.

Scion-wood of conifers prepared for grafting. These are best when used fresh but can be stored for about 10 days in a cold store.

5 Bench Grafting

IN RESEARCHING THIS BOOK, SEVERAL growers who specialize in grafting have given me a fantastic amount of help in explaining their approach to grafting. One thing became clear to me very quickly: there is no single method to use when grafting a plant. What all the growers had in common was that they have successful nurseries that grow high quality plants that are in demand and make a profit for the business. The other thing they had in common was that each grower had their own approach to producing their plants. Certainly, they followed the basic principles required to achieve a successful graft union, but after that, variations in the type of graft they used, the timing of the graft, the rootstock, and the aftercare meant that they were quite different from one another. And "different" means just that—not better, preferred, or correct, just different. There is no point, therefore, in trying to provide a grafting technique to use for every species. The aim of this chapter is to look at why the different methods are used, how they successfully produce grafted plants, and to provide some case studies to act as a guide for anyone wishing to start grafting and help them select a method to use.

Elements of bench grafting

Bench grafting refers to any type of graft that is made in a potting shed or similar area "at a bench" rather than in the field. It is a long-established technique for plants that need some protection or heat to form a union and are not suitable for budding in the field. It is also used as an alternative to field budding to produce plants for particular purposes or form.

Bench grafting is primarily used in four situations. The first is for evergreen species like conifers and rhododendron that are not suitable for budding in the summer but require warm conditions

for the union to form. The second situation is with deciduous species that may not produce suitable bud material or be hardy enough to be grafted in the field, for example, *Betula pendula*, *Fagus sylvatica*, *Wisteria* species, or *Acer palmatum* cultivars. In the third situation, bench grafting is used to produce top-worked plants as patio plants or as container-grown, light standard trees for sale through garden centres.

The fourth application of bench grafting is when species that are traditionally budded in the field are needed for particular uses. One such example is when apple trees are grown for the Dutch knip-boom orchard system. Knip-boom (literally cut knee in Dutch) produces a seven-branched well-rooted tree that can give early fruit production of 13–20 tonnes per hectare of fruit in the second season after planting. Bench grafting can provide the required tree for planting in

Second year knip-boom trees showing the well-branched tree that enables early fruit production once in the orchard.

two years. Bareroot rootstocks are grafted in late winter and planted out in the field in spring. The leader, which puts on about 80 cm of growth in the first season, is then headed back to 40–60 cm the following winter. Cutting the leader like this will produce about seven feathers, resulting in a strong tree capable of earlier cropping.

Another occasion when bench grafting is preferred for a species that is traditionally budded in the field is when an interstock is required. Bench grafting in this instance allows the three plant parts to be joined together prior to planting. This would be difficult to achieve if budding in the field.

One final situation where bench grafting is superior to budding in the field is with M27 rootstocks. M27 rootstocks may produce better plants if bench grafted and container-grown for a season prior to planting in the field. This is because M27 does not always produce vigorous enough growth when planted directly into the field.

In addition to the type of graft used, the other major variations when planning a bench grafting programme are timing (late summer or early spring), rootstock (bareroot or container-grown), and control of environment, especially ambient or additional heat. Some decisions will depend on the species being grown but often the type of graft used will depend on location, preference, or facilities available.

Types of graft: Side or apical

Two types of graft are used depending on the species and preference of the grafter: side grafts that retain the top of the rootstock while the union is formed, and apical grafts that attach the scion to the top of the rootstock that has been reduced to the required height. Both types of grafts can be used for deciduous and evergreen broadleaf shrubs and trees, but for evergreen conifers, a side graft seems to be a requirement

The side veneer is the most common side graft

used along with the modified side veneer (or side wedge), which maintains a flap on the rootstock to provide a union on both sides of the scion. The side veneer graft requires fewer cuts and is probably an easier graft to prepare, but the modified form gives greater cambial contact and may be stronger initially after union.

The main apical grafts used are the cleft, wedge, saddle, whip, and whip-and-tongue. The cleft graft is a straightforward graft in which a cut is made down through the centre of the rootstock and a scion inserted that has been cut on two sides. This type of graft helps to apply pressure on the scion, but care has to be taken not to split the rootstock when carrying out the cut. It also gives a large area of cambial contact and is useful if the scion diameter is smaller than the rootstock so that the graft can only be applied to one side of the rootstock cut.

The **wedge graft** is very similar to the modified side veneer graft but with the top of the rootstock removed. This graft is easily prepared if the propagator is familiar with the modified side veneer graft. A wedge graft gives a large area of cambial contact. The flap does not apply pressure to the scion like the cleft, and some grafters believe it is a little weak immediately after the union forms.

The **saddle graft** is the reverse of the first two grafts in that the scion fits over the rootstock rather than into it. It is used where larger diameter grafts are prepared, such as with rhododendron, or where harder wood is used, like in larger diameter apple rootstocks.

The **whip graft** requires only two cuts. It is a straightforward graft to prepare when the diameters of the rootstock and scion are matched. If the diameter of the scion is slightly less than the rootstock, then the depth of rootstock cut can be adjusted down the side of the rootstock so the diameter matches the scion. The whip graft does need to be tied-in by the grafter and practice is required to hold the cut together while tying.

The **whip-and-tongue graft** adds an extra vertical cut into the scion and rootstock whip cuts. As a result, it is easier to hold the graft together when tying-in, it helps to apply pressure to the union, and it increases the area of cambial contact. It does mean making extra cuts compared to the whip graft. A second person can, however, work with the grafter to tie-in the prepared grafts.

The use of a whip or whip-and-tongue graft often comes down to the propagator's preference, though for a few species, the type of graft used may be important. A whip-and-tongue is preferred for *Aesculus* and *Gleditsia* grafts as they are slow to form graft union and it is important to ensure there is no movement at the union during this time. With *Juglans regia*, a species with a very pithy stem, whip-and-tongue graft causes too much cell damage, so a whip graft gives better results.

Rootstocks: Bareroot or container-grown

It bears repeating that the rootstock is critical to making a successful graft. It must be the correct species or cultivar with the correct diameter stem. It must have a good volume of healthy fibrous roots, and active cell division must occur at the time of grafting. Summer grafting uses container-grown plants, but in winter the rootstocks can be either bareroot or container-grown. Bareroot plants are generally less expensive as they will either be lifted from the field or delivered dormant prior to grafting.

Container-grown plants are usually grown on by the nursery for a year prior to grafting. However, cell-grown plants are used by some growers for conifers and other plants which are bought in the autumn and grafted the next spring without growing on for a season (as long as they have the correct stem diameter). Although container plants are more expensive to produce, they have the advantage of established roots that will be actively growing at the time of grafting.

With many plants, using either bareroot or container-grown plants will give successful grafts. With some, however, the percentage take may be lower with bare rootstock and the subsequent growth should be considered. Since the bareroot plants need to be potted on after grafting and the roots then need to grow into the medium of the container, the growth can be more variable than is usually found with the established container-grown rootstocks. If this variation in growth means that a significant proportion of the plants do not make the quality specification by the end of the growing season, then it is worth considering using container-grown rootstocks in the future.

Timing: Early spring or late summer

As far as the rootstocks are concerned, the timing is mainly about active cell division and sap flow. Cell division must be occurring within the rootstock so that the new vascular cambium system can form. At the same time, the flow of sap needs to be controlled so that it is not excessive. Excessive sap flow can lead to the pressure from the sap pushing the graft apart and causing the union to fail. It can also cause sap to leak at the union and cause disease problems and graft failure.

In spring, the sap is rising as the new season's growth begins. Bareroot plants will be taken directly from a cold store or a temporary planting area outside and grafted while they are dormant but with the sap just starting to rise. Container-grown plants are brought into a greenhouse over winter, or a few weeks prior to grafting, so that root growth begins with about 6.5 mm of fresh roots showing when the graft is undertaken. Watering is critical to control sap flow and must be carefully managed so the rootstock is not overwatered.

In late summer, sap flow decreases as the season's extension growth ends and active growth starts to reduce towards winter. The sap flow through the plant is naturally declining at this time and it may be possible to keep the rootstocks well watered without getting excessive sap flow. However, often the rootstocks will be dried back and then watered sparingly prior to grafting to control sap flow.

Whether the bench graft is carried out in the late summer or early spring will often depend on the preference of the grower. Some growers carry out 80 percent of their grafting in early spring as other tasks on the nursery mean that the summer is a much busier time and difficult to fit in grafting. Other growers prefer to complete most of their grafting in late summer and mainly graft in the early spring because there is simply not the time to complete all the grafting in summer.

One advantage of late summer grafting is that warmer conditions give a fast take and the plant is then ready to grow the next spring. Although this may not be critical for many species, some, like *Corylus avellana* 'Contorta' (contorted hazel), put on better growth by the end of the first growing season if they are grafted in summer rather than spring. Timing is important with contorted hazel because it is quite slow growing. In contrast, the preferred time of grafting for *Fagus sylvatica* cultivars is spring; these plants seem to do better at this time of year if grafted and placed on base heat. This again illustrates quite different, but successful, approaches that are mainly due to staff available and the experience of the propagator.

Whether grafting at the start or end of the season, the scion buds must be, and remain, dormant until the graft union has formed. There are a number of requirements for the buds of temperate plants to grow. They have to undergo a period of cold chill after they have formed before they will break into growth. Temperate plants have evolved this requirement so that there is less chance of shoots starting to grow when they are likely to be damaged by frosts and inclement weather. In the spring, buds will develop depending on

air temperature and the influence of the roots. If the temperature is high enough, then scion growth can occur even if the graft union has not formed. Once the union is formed, nutrients and growth-promoting substances translocate up the stem and promote growth.

Summer grafting has the advantage of cell division still occurring in the actively growing rootstock, but the scion buds will not grow until they have received a period of cold conditioning. The timing of the graft will require the new season's extension growth to have ended and the buds to have matured. The scion-wood must also be firm enough, that is, have the correct C:N ratio indicated by the firmness of the stem. At this point, the active growth of the rootstock will be slowing so that sap flow is reduced and can be controlled. Grafting cannot continue too late in the season since the temperatures and light levels have to be high enough for the graft union to form successfully. This means that summer grafting normally is carried out from late summer through early fall. Some species that put on only one flush of growth in a season, like *Acer palmatum*, meet the requirements early in the season and can be grafted in early summer.

The other consideration with summer grafting is that limited waxing is possible. Since even deciduous scions retain leaves at this time, it is only possible to brush wax over the union; the scion cannot be waxed. Grafts must be kept in an atmosphere of high humidity to prevent desiccation, usually by keeping them under a polythene tent. However, for summer grafts contact, polythene may be preferred above the grafts. (See Temperature for further discussion on this.)

Late-winter grafting has the advantage that, once formed, the new graft will be going into active growth in the spring. One problem with grafting at this time is that the temperatures must be high enough for active cell division to occur within the rootstock, but not so high that the scion buds will break before the union has formed and cause the graft to fail. One way to reduce the chance of early bud growth from the scion is to collect the bud-wood of deciduous material in midwinter, and cold store it between 0 and 3°C to ensure it remains fully dormant prior to grafting in late winter or early spring.

With apical grafts at this time, moisture can be retained around the union by waxing or using polythene ties, and a high humid atmosphere is not required. When side grafts are used, with either deciduous or evergreen plants, then waxing is not possible over the whole graft and so a high humidity must be maintained by the use of polythene. In the lower light levels and temperatures of the spring, it has been found that contact polythene gives poorer results than a tent above the grafts. This is principally due to higher levels of *Botrytis* occurring under the contact polythene.

Temperature: Ambient or additional heat

The lowest temperature that can be used is the minimum for cell division to occur within the rootstock. In the case of many Rosaceae species like *Malus*, *Prunus*, and *Sorbus*, any temperature above 0°C will see some cell division occurring, although the optimum temperature will be around 25°C. This means that the graft can be quite early when ambient temperatures are still low and the aftercare is relatively straightforward. The union will form slowly at these temperatures but successful grafts will be achieved. *Juglans regia* (walnut), a difficult species to graft, will not graft at the lower temperatures and requires an optimum temperature of about 32°C.

Newly made summer grafts, or side grafts in early spring, must be kept under polythene covers while the graft union forms. In this case, temperature management is critical and must be care-

fully carried out. An increase in temperature by one degree can lead to a 3 percent drop in relative humidity; it does not require a large rise in temperature to cause desiccation or scorch to the scion material in a graft.

Another problem that can arise is early scion growth before the union has formed properly. Scion growth can occur at temperatures just above those required for the union to form, but below those that would cause desiccation or scorch. At these temperatures, scion buds will start to grow before the union has formed and then collapse because they are not supported by the roots.

If there is too much shade, then light levels will fall below that required for photosynthesis and the graft union will fail. Careful shading is therefore required to ensure enough light for photosynthesis but not excess to cause a build up of heat and reduced humidity under the polythene.

Early spring grafting—cold callusing method

Cold callus grafting refers to a slow take of the graft union at low temperatures. For many Rosaceae species like apples and rowan (*Sorbus aucuparia*), this technique is very successful. The plants form a graft union at temperatures above 0°C, and bareroot rootstocks are generally used for this type of material. After the grafts are prepared, they can be treated in one of several ways. They can be potted straight away and stood down in a sheltered area or unheated greenhouse. They can be placed together in crates with the roots covered in a suitable medium to prevent them drying out and set in a sheltered site or unheated greenhouse. This treatment reduces the amount of space required while the graft is forming and allows the propagator to pot up the plants later when time allows. Alternatively, they can be planted out into the field immediately if they are to be field-grown rather than container-grown.

Some non-Rosaceae species can also be grafted cold from bare rootstock. *Wisteria*, for example, can be grafted successfully in midspring with the rootstocks potted up after grafting and placed under a low polythene tent in a greenhouse. At this time, no additional heat is required although other growers will graft *Wisteria* earlier and place on a hot-pipe (discussed later).

Other trees and shrubs are a little more demanding of the environment they require to form a union. For example, *Betula*, *Cercidiphyllum*, and

Syringa grafts should be carried out using container-grown rootstocks and, after grafting, these are kept in a greenhouse at three to four degrees above the outside (ambient) temperature. Although this looks very similar to bareroot grafting, rootstock preparation, temperature control, and water management are much more critical with this method since these species have a slightly higher minimum temperature than the Rosaceae species.

In Australia, successful grafting of native species like *Corymbia*, *Grevillea*, *Hakea*, and *Brachychiton* only began in the 1960s. *Corymbia ficifolia*, for example, produces red, orange, scarlet, and pink

Table 4. Possible shade levels for grafts under low polythene tent. This will provide light for photosynthesis but keep temperatures low enough to avoid scion bud break.

	Number of layers of 40% shade cloth required							
	Latitude 50°N (Portsmouth, UK; Regina, Canada)		Latitude 45°N (Milan, Italy; Portland, USA)		Latitude 40°N (Madrid, Spain; New York City, USA)		Latitude 35°N (Nicosia, Cyprus; Memphis, USA)	
MONTH	BRIGHT DAY	DULL DAY	BRIGHT DAY	DULL DAY	BRIGHT DAY	DULL DAY	BRIGHT DAY	DULL DAY
January	0	0	0	0	1	0	1	0
February	1	0	1	0	1	0	1	0
March	2	0	2	0	2	0	2	0
April	3	1	3	1	3	1	3	1
May	3	1	3	1	3	1	3	1
June	3	1	3	1	3	1	3	1
July	3	1	3	1	3	1	3	1
August	3	1	3	1	3	1	3	1
September	2	0	2	0	2	0	2	0
October	1	0	1	0	1	0	1	0
November	0	0	0	0	1	0	1	0
December	0	0	0	0	1	0	1	0

Source: Adapted from K. Loach (1981).
1 layer of 40% shade cloth = 40% reduction in light transmission
2 layers of 40% shade cloth = 64% reduction in light transmission
3 layers of 40% shade cloth = 78% reduction in light transmission

Cold callus grafting from bare rootstock

Bareroot rootstocks kept outdoors in a temporary planting (or 'sheughed' in to use a good Scots phrase). This is best in a site that does not get direct sunshine.

Quince A rootstocks waiting final preparation prior to grafting.

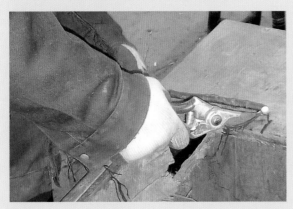

Rootstocks are cut to length.

The stem of the rootstock is cleaned prior to grafting. It is important that dirt does not get on the cut surfaces and form a barrier to the union forming properly.

Preparation of the scions. Bud-wood of the correct thickness and firmness with healthy, mature buds is selected and cut to three buds. Since the interstock will be grafted top and bottom, a longer piece is cut for grafting, about the length of a pair of secateurs.

The whip cut being made on 'Commice' used as an interstock between Quince A rootstock and 'Conference' and 'Williams' pears.

The whip cut could be joined with the rootstock and this is common practice. In this case a tongue is added to the whip to hold the graft together prior to tying. The process involves one person grafting and two tying-in.

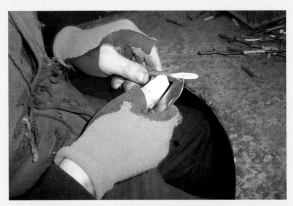

The scion tongue is cut in the lower third of the whip cut. When the rootstock tongue is cut in the top third, this allows the scion and stock to be held together and the cut surfaces to match.

The whole of the scion and cut surfaces being dipped in molten grafting wax. The wax should be between 60 and 70°C.

The graft being joined to the rootstock. Note that the 'Conference' scion has already been grafted and sealed onto the interstock.

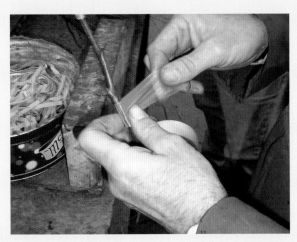

The interstock is tied to the rootstock with Buddy Tape. This not only holds the graft together but also seals the cuts, as it will not be dipped in wax. Do not overwrap the tape or it will not degrade and thus the propagator will have to remove it, which is difficult as it is self-adhesive. Buddy Tape or polythene tape can be used for grafts not requiring an interstock. They seal the wound as well as holding the graft together. Petroleum jelly placed on the top cut of the scion prevents desiccation.

Cold callus grafting from container-grown rootstock

Prior to grafting, the container-grown rootstock is started into growth. At least 6.5 mm of fresh roots should develop prior to grafting. It is very important to control the sap flow through the stem by watering sparingly.

The rootstock cut is made by cutting towards the body, with the container supported at an angle towards the propagator.

The width of the scion and rootstock cuts should match. If the scion-wood has a slightly narrower diameter than the rootstock, then make a shallower cut up the side of the rootstock so the two widths will match.

The whip graft is secured by a rubber tie that will then be waxed. Polythene tape can also be used and does not require waxing as long as all the cut surfaces are sealed by overlapping the tape. The top of the scion is sealed with some petroleum jelly.

When waxing a container-grown plant, make a cover out of cardboard to put over the medium and prevent it falling out.

Completed graft. Note the depth of cut in the rootstock.

The cuts match precisely giving excellent close cambial contact that will help the union form quickly even though it is grafted cold. This precision will only be achieved with a lot of practice. Note the 'church window' on the scion at the top of the graft. This ensures the cambium is running parallel when the scion and rootstock actually come in contact.

Completed tunnel of grafts. Note the uniformity of height and graft union.

flowers but is difficult to propagate vegetatively. A method of grafting the species is described below although it has been further refined by growers to achieve a high percentage of successful grafts. Seedling rootstocks are used and these should be vigorous and of the correct diameter. The grafts are made in late summer, when the sap flow is slowing down. Scion-wood should be mature and of the current season's growth. The scion leaves are retained but reduced in size, although some growers remove the leaves and just retain the petioles.

An apical graft is used but two healthy leaves should be retained on the rootstock below the graft union. It is important to get a precise match of scion and rootstock cuts so that there is no sap leakage at the union. The graft is tied in with polythene tape or Buddy Tape, and petroleum jelly is used to seal cut surfaces like the end of the petioles. The grafts are then kept in a greenhouse and the medium is prevented from drying out.

Case study: Bareroot rootstocks of Rosaceae species

As previously noted, bench grafting can be used for fruit trees like apples (*Malus*) and pears (*Pyrus*) as well as ornamental trees like rowans (*Sorbus*) and ornamental cherries (*Prunus*). The rootstocks should be obtained in the autumn prior to grafting in the late winter. Since they will not put on any growth until after they have been grafted, a diameter of 9–11 mm is best to be used. One-year rootstocks of apples will often provide

Grafted *Corymbia ficifolia* 'Calypso Queen' at Humphris Nursery, Victoria, Australia. Note the two leaves retained on the rootstock beneath the graft union.

Close up of the graft union of a weeping eucalypt at Bywong Nursery, New South Wales, Australia.

the required girth, but it may be easier to obtain two-year-old stock of this girth. Pear, plum, and cherry rootstocks will be two years old. Rootstocks of *Sorbus aucuparia* will be seed-raised, and an alternative to using bareroot plants would be to use cell-raised plants. The rootstocks can be stored in a cold store until required. Alternatively, they can be temporarily planted in a plunge bed outdoors and covered with a layer of bark or similar material to protect roots from drying out. The site should be sheltered from direct sunlight so that the rootstocks do not start to grow actively in the spring before they are grafted.

Scion material should be collected in midwinter when the buds are still fully dormant. The scion-wood is wrapped in plastic to prevent it from drying out and labelled inside and out to avoid mistakes in labelling the final plants. It is then stored between 0 and 5°C to ensure the buds remain dormant. The scion material should match the rootstock diameter with well-formed buds. Any thinner, bendy material should be discarded.

Prior to starting to graft, the rootstocks and scions should be prepared. If they are to be potted after grafting, then the roots are trimmed to a size that will fit into a 3-litre tall container. Quince roots tend to dry out quickly so may be soaked after root pruning. It is also important to pot these up quickly after grafting. Grafting should take place in a cool environment and some propagators like to wet the floor to create a humid environment to ensure that roots do not dry out.

The stem of the rootstock is then cleaned and cut to length. Where vigour control of the stock is important, rootstocks should be cut to 20 cm long. This will prevent future rooting of the scion that would negate the benefits of the rootstock. For ornamental grafts, where the rootstock is only for propagation purposes, the length of the fist and thumb is used as a guide for cutting the rootstock—approximately 12 cm. Some grafters prefer

to cut the rootstock above a node to reduce the risk of dieback in the same way that you should cut to a node when pruning shrubs and trees in the garden. It is not essential to cut to a node, however. The scion is cut to three buds or to a length that is easy to hold. If an interstock is to be used, this should be cut to the length of the handle of secateurs (about 14 cm) as two cuts will be made on this material.

Some people prefer to use a knife to make the final cuts to the scion and rootstock, as it gives a cleaner cut than secateurs. It is important to ensure that the buds on the scion material are healthy. If there is any staining on the outside, or vascular staining on the inside after cutting, then the plant material should be discarded. The preparation of the whip or whip-and-tongue graft depends on your preference or the requirements of specific species.

Whatever material is used for tying, it is important that the tie be applied tightly enough to hold the graft firmly together and to apply pressure to induce the differentiation of the cells into the new vascular cambium system. It is most important to ensure the ties are secure at the top and bottom of the graft so that these areas seal and do not leave a weak point.

If using rubber ties, they should be 200 mm long by 6 mm wide. It is not critical to overlap the strips, but the graft should be wrapped closely down the whole length. The wrapped graft is dipped in wax to seal the cut surfaces and prevent moisture loss. Use grafting wax that has a relatively low melting point. Keep the temperature of the wax between 60 and 70°C. Wax below 60°C will be applied too thickly and can be brittle, while wax above 70°C can damage the wood of the rootstock. Although candle wax has a higher melting point than is ideal, it can be used: give the graft a quick dip in and out of the wax to avoid cell damage to the wood. Rubber ties do not need to be removed once the grafted plant starts to grow and the stem girth expands, as the rubber will go brittle and break away rather than restrict the stem expansion.

Alternative materials that do not need to wax are polythene grafting tape and Buddy Tape. These materials need to be overlapped to cover the entire cut surface. Since the polythene is not biodegradable, it does require to be removed by the grafter, unlike the rubber ties. The timing of removal of the tie is critical. If it is too soon, the graft union will be too weak and the scion can be easily knocked off the rootstock. If it left on too long, then the tie can restrict the growth of the stem. Initially, the callus around the graft union will appear white; it then goes green and finally starts to go brown. At this point, it should be possible to remove the tie safely.

Buddy Tape is a biodegradable material, but if it is excessively overlapped then this will not occur and it will have to be removed. This can be time consuming since it becomes self-adhesive when stretched to tie, and so does not easily unwrap once the graft union has formed. It may be better to use only half a patch of the tie and only wrap the graft with three turns of the tie. This should enable the tape to degrade enough so that it does not need to be removed by the grafter. Another disadvantage of Buddy Tape is that it will not apply as much pressure as the rubber or polythene ties. This is not a problem if the cuts match well, but if a cut is not made precisely, the ability to tie-in tightly with rubber or polythene is important.

For knip-boom trained apples, the first year's growth will be cut back to 60 cm, which can easily be achieved after planting out in the field. For grafts that are going to be grown on in containers for the garden centre market, then maiden growth

After grafting, these grafts have been placed in trays and placed in a cold glasshouse for the union to form. They will be potted up later when there is more time and space.

of around 2 m will be required so that a light head can be formed in the second year. In this case, growing on the plant in an unheated greenhouse will usually achieve these requirements. This assumes that the graft is well made so that there is good cambial contact and the union forms relatively quickly.

I have found that with students who are just learning to graft and have not had a lot of practice, cold callusing can give very variable results. I use the hot-pipe callus system for the likes of apples, at a temperature of about 22°C. This has given a better graft take over the last few years, and, since the union forms more quickly, the subsequent growth usually gets to 1.5–2.0 metres in a season.

Even with experienced grafters, cold callusing onto bare roots can sometimes produce plants of variable height after one season that may not achieve the required grade. Ornamental pear

Maiden growth of rowan (*Sorbus aucuparia*) in early autumn after being cold grafted in the previous winter.

In this case, the grafted pear trees have been potted immediately after grafting and are going to a cold greenhouse to be stood down. This may reduce double handling if there is space on the nursery for the containers.

cultivars of *Pyrus calleryana* can give variable maiden growth the season after grafting. This seems to be due to variation in the rootstocks establishing once potted up and starting to grow. This problem has been overcome by using pot-grown rootstocks that are already established and initiate growth more quickly and evenly after grafting.

Case study: Container-grown rootstocks of *Betula*

In some areas like Oregon on the U.S. West Coast, the growing conditions allow grafters to obtain a good diameter of scion growth at the correct time to chip bud *Betula*. More commonly, however, chip budding is not feasible and bench grafting is used. The sap rises early in the season with birch and can be a problem to control. For this reason, a side veneer graft is commonly used and this method of grafting is described for conifers. However, it is possible to use a cold callusing method for birch and many other shrub species, such as *Cercis*, *Cercidiphyllum*, *Cytisus*, *Laburnum*, *Robinia*, *Syringa*, and *Viburnum*. For these species, using a whip graft that does not require to be headed back is quicker and easier to prepare than the side veneer graft, and it produces a strong union, with callus formation over the cut on the rootstock.

Although the whip graft appears to be a straightforward technique, it requires great attention to detail to give viable results. The graft needs to be very well prepared to match the diameters of scion and rootstock, giving close cambial contact that will form the union rapidly even under relatively cool conditions. The rootstock has to be carefully prepared and the temperatures managed correctly.

One-year-old seedlings of *Betula pendula* with a girth of 4–6 mm are bought in, potted up into 1-litre-deep pots, and grown on for a season to produce well-established plants of 6–8 mm girth.

The rootstocks are brought into an unheated greenhouse in the autumn to dry off and are kept just moist to the touch. The tunnels are ventilated on sunny days to prevent too early growth. Scion material (true-to-type, healthy, good girth, and firm with healthy buds) is collected in early winter and wrapped in strong plastic bags and stored at 0–5°C. The scions can be stored for around two months in a cold shed and longer if a cold store is used. The whip graft is prepared, waxed, and stood down in a greenhouse. The aim is to keep the air temperature around the grafts in the greenhouse 3 to 4 degrees above the outside temperature. Once the temperature outside reaches 10°C, then the greenhouse is ventilated. This provides temperatures above the minimum for the union to form, but not too high that early growth of the scion

A one-year-old *Betula* graft.

occurs. If ventilated correctly, the scion growth should not be more than two to three weeks ahead of growth on plants outdoors.

The management of the rootstocks is also critical. At the time of grafting, the medium in which the rootstocks are growing should be just moist to the touch and the rootstocks should be showing at least 6.5 mm of fresh white root growth. Watering should be done sparingly for several weeks, and only once the scion growth is 20–30 cm tall can water be applied more freely.

Alternative techniques can be used for birch. After grafting, they can be put on a callus hot-pipe. This should not be too hot; 16–23°C is recommended. This will achieve a faster graft union and the grafted plants can then be placed in a greenhouse to encourage early growth. Excessive sap flow can be a problem with birch on the hot-pipe, as it is a plant that leafs out early in the spring. For this reason, side grafts are still popular for this plant so that the rootstock acts as a sapdraw. (See the section on conifers for more information on managing this type of graft.) Perhaps for this reason, the lower hot-pipe temperature of 16°C would be preferred.

Early spring grafting—hot callusing method

The application of artificial heat from a boiler (or other heat source) to a grafted plant or part of that plant can be described as hot callusing. This method has the advantage of forming the graft union more quickly than cold callusing as it provides a temperature nearer to optimum for the graft unions to form. This in turn leads to earlier growth of the scion and usually greater maiden growth. Hot callusing can be used on any species that can be grafted by cold callus methods, but is a requirement for other species that require higher temperatures to form a union in the late winter.

Hot callusing has been tried by simply putting the grafted plants into a heated greenhouse, but the higher temperatures that stimulate callus formation can also cause the scion buds to break dormancy and leaf out. When this occurs before the union has properly formed, the growing scion cannot take up enough water and the new growth collapses. The traditional hot callus system to use, where deciduous species required higher temperatures, was the side veneer technique—the same process undertaken for conifers. Under the high humidity of the polythene tent, and as long as the air temperature does not rise above 18°C, the warmer conditions allow the union to form but keep the scion buds dormant. It is only once the graft union has formed and the scion has connected to the rootstock that the influence of the rootstock will promote bud break. It is important to remove these grafts from the tent as soon as the leaf has started to show. The rootstock is then headed back over the following months as described for conifers. This type of side graft also makes water management slightly easier than if an apical graft is used, since the rootstock will act as a sap draw reducing the pressure of sap at the graft union.

To overcome the problem of early scion bud break, heat must be focused where it is required—at the graft union. With heat directed at the graft union, the scion buds and rootstock remain cold and dormant, while the actual graft union is at the optimum heat to speed up the graft formation. This system enables an easier apical graft to be used and simplifies the aftercare but gets the benefits of higher grafting temperatures.

Hot-pipe callus system

In the 1970s, Harry Lagerstedt, an associate professor at Oregon State University, started to work on a method of applying heat to the required part of the graft. By 1983 he had a practical system and was awarded US Patent 4,383,390 "Method and apparatus for hot-callusing graft unions" (the patent lapsed in 1992). This technique not only gave as good results as the side veneer graft and was easier to manage, but also the graft take often was more successful. It has now become widely used in place of side veneer grafts for many deciduous species in the winter.

The hot-pipe callus system enables the graft union to receive the optimum temperature for callus formation but keeps the scion buds dormant and roots cool. The rootstock can therefore be cut back to size and the scion attached by a whip graft, or similar, as a sap draw is no longer required. A

polythene tent and high humidity are also not required and so diseases like *Botrytis* are less likely to be a problem. Since the scion and rootstock are dormant there is very little translocation and managing the watering is less critical. Bareroots just need to be covered with a suitable absorbent material and kept moist. Pot-grown rootstocks can also be used but do not need to be in active growth prior to grafting. Timing is also less critical for the removal from the hot-pipe. Three weeks is the usual time required although the optimum time depends on individual species. It is also possible to start grafting earlier, but plants should be placed back in the cold store once the callus has

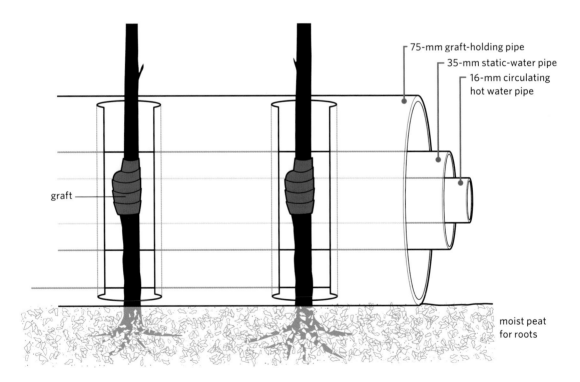

Figure 5-1. The inner pipe has flowing warm water from a boiler. The second pipe is filled with water that is heated by the warm water flowing from the boiler. This was an innovation to get temperatures that are more consistent; most systems do not use the pipe with the static water. The outer pipe has slots cut in it to hold the graft unions. The air in this pipe is warmed by the water pipes and prevents the grafts encountering direct heat. Insulation surrounds the pipe work to reduce heat loss, and the roots, if not in containers, are covered in a moist medium to prevent them drying out.

formed. They will then remain dormant until the weather is suitable to plant out or stand down.

There are many designs of callus hot-pipes but the principle is the same. The graft union has to be placed in warm air while the root and scion buds remain in cool air. Warm water systems are commonly used with two pipes—an inner one circulating warm water which warms the air in the outer pipe. Notches are cut in the outer pipe that hold the graft union in the warm air of the large gauge pipe. It is important that the graft does not come into direct contact with the heat source.

Frank P. Matthews' nursery found that this design gave fluctuations in the air temperature, so it redesigned the system to put the circulating warm water pipe within a larger diameter pipe containing static water (Figure 5-1). The new arrangement reduces temperature fluctuations and enables the optimum temperature to be provided for each species being grafted. Insulation around the outer pipe keeps the heat within the pipe while keeping the roots and buds cool. The roots are kept in moist peat or similar moisture-retentive material, or are in containers and the scion remains at ambient temperature. The nursery's hot-pipe has been built in an old cold

The latest design of the hot-pipe system at Frank P. Matthews' nursery uses electric heating cables in place of water and is built up of units about the size of pallets, making it very flexible.

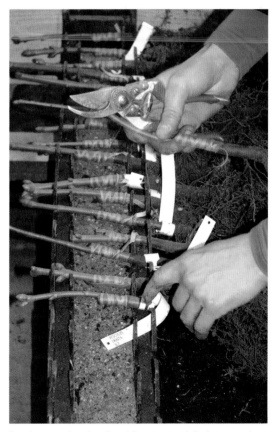

Sand covers the electric cables to spread the heat evenly and the graft union is placed in slots over the sand.

store which remains cool and does not heat up rapidly in spring sunshine. A third generation of hot-pipe has now been built at Matthews' nursery and it uses electric cable to heat the air to provide a more flexible system with accurate temperature control.

The callus hot-pipe method can be used for all deciduous grafts but, if space is limited, the Rosaceae species like apples and rowan can be grafted cold. This is preferable for apple trees being grown for knip-boom production, and which are planted outdoors after grafting, as any more than 1 m growth will be wasted. The faster union on a hot-pipe will give earlier scion growth in the spring and greater extension growth of the maiden in the first season. This may be preferable to reduce production times of either container- or field-grown trees.

Although it might seem possible to also use the hot-pipe for conifers, this has proven to be unsuccessful. It is difficult to maintain the humidity around the hot-pipe and it has been found that the lower scion buds are within the heated area and grow too early, causing establishment problems.

Another situation where it is better not to use the hot-pipe is when a scion has a nurse bud, since the bud would be within the warm pipe and initiate early scion growth at that point.

An insulated board is placed over the grafts to keep the heat around the graft union.

After three weeks, the callus is well formed and the union has occurred.

Case study: Hot-pipe callus grafting of deciduous *Magnolia* species and cultivars

Besides deciduous magnolias, a wide range of deciduous shrubs and trees can be grafted using the callus hot-pipe, among them *Aesculus*, *Amelanchier*, *Betula*, *Carpinus*, *Castanea*, *Cornus*, *Gleditsia*, and *Quercus*. This technique has played an important role in simplifying the grafting process and improving the successful take of these species. Hot-pipe callusing can probably be used for all deciduous shrubs and trees but, as discussed elsewhere, it is not usually required for apples, and some plants do better when summer grafted. Contorted hazel (*Corylus avellana* 'Contorta') will graft as well in early spring as in late summer, but the subsequent maiden growth is better from summer-grafted rather than spring-grafted plants. Either bareroot or container-grown rootstocks can be used on the hot-pipe. Again, this often comes down to the grower's preference, but sometimes there may be a difference in percentage graft take or subsequent growth of the scion.

The genus *Magnolia* is divided into two subgenera *Magnolia* and *Yulania*. For *M. ×wiesneri* and other species in subgenus *Magnolia*, a suitable rootstock is *M. hypoleuca* from seed. Most magnolias commonly grafted belong to subgenus *Yulania* and many are compatible with one another. When selecting a rootstock for these magnolias, it is more important to match vigour between scion and rootstock since there are few incompatibility issues. *Magnolia campbellii*, for example, is tree-sized and would not be a suitable rootstock for the shrubby *M. stellata*. In this case, *M. kobus* would be the preferred rootstock.

Magnolias can be grafted in late winter or chip budded in the summer. Grafting in late winter involves a straightforward technique that produces a good-sized plant after the first season. If scion material is limited, then chip budding has the advantage over grafting in that more plants can be produced from the same amount of material. A compromise is to whip graft with just two buds. Since a single leader will be grown once the graft has taken, it may even be possible to whip graft using a single bud. Some growers, however, also believe that a chip bud gives a better union and a better-shaped plant.

Magnolia hypoleuca and *M. kobus* are both raised from seed. For late winter grafting, two-year-old bareroot seedlings with a 5–6 mm diameter can be used, or rootstocks grown for a season in P9 pots. With budding, the rootstocks are established in 1-litre pots during the season prior to budding.

Bareroot rootstocks can be used for grafting in the late winter using a splice graft and placing on a

Growth on a *Magnolia* grafted in late winter using the hot-pipe system. A two-bud scion was used and one bud was retained after the union formed to produce the stem.

Taking a chip bud in summer. Nick Dunn, of Frank P. Matthews' nursery, learned T-budding originally and so has adapted this technique for chip budding.

The bottom cut on the chip bud is prepared after the bud has been removed, unlike the conventional technique.

Cutting the rootstock to take the chip bud.

Using Buddy Tape to tie-in the bud. Only half a patch was used and only three turns were used. If the tape is overwrapped, it will not degrade. The propagator will have to remove it, with difficulty.

hot-pipe at 24°C for three weeks. Using container-grown rootstocks with *Magnolia* improves take by approximately 20 percent. Once the graft union has formed on the hot-pipe, the graft is put into a greenhouse where frost protection is available. When the scion starts to grow, a single bud—usually the top bud—is selected to be grown on. A straight stem is produced by tying it to a cane. All the other buds on the scion, particularly on the rootstock, are removed by rubbing off when new growth is short and soft.

Chip budding can be carried out in the late summer on container-grown rootstocks. As with other summer grafting, the scion-wood must be firm but not too woody, with well-formed buds. The leaves are removed with a knife or secateurs leaving no more than 1 cm of the leaf petiole. Preparing the chip bud is described in full in chapter 6. The bud is applied about 10 cm from the base of the rootstock and tied in, leaving the bud uncovered. It is important not to trap the leaf petiole under the tie as, once it abscises, it can rot the chip. This can also be avoided by pulling off the leaf petiole entirely when preparing the bud-stick.

Once budded, the grafts are put under a polythene tent. It is important to keep the atmosphere under the tent as cool as possible by shading. The callus will form rapidly, within about two weeks, at which point the graft can be weaned from the polythene tunnel into a drier atmosphere. The following late winter, the rootstock is headed back, the grafts are potted on into 5-litre containers, and the new growth tied to a cane.

Late summer grafting

In late summer and early autumn, the natural ambient temperature is still high and so the temperature inside a greenhouse can be maintained around the optimum for the graft union to form.

Either side or apical grafts can be used at this time. For both types of graft, managing the humidity levels, usually under polythene, is necessary since the plants still have leaves and cannot be waxed beyond the graft cuts. It is, therefore, critical to manage light levels and temperatures by shading and ventilation to achieve successful unions.

In early spring, *Botrytis* can be a significant problem when side-grafted plants are kept under polythene. This is because light levels and temperatures are still quite low and, if humidity should become high, the fungal disease can quickly become established. At this time of year, grafts kept in a low polythene tent that leaves air space around them will be less likely to have *Botrytis* problems than grafts that make contact with their polythene cover. In late summer, however, with higher temperatures and light levels, the risk of fluctuations in humidity and temperature are greater than disease problems. It is therefore preferable to have the polythene in contact with the grafts, to reduce the air space around the grafts and any fluctuation in humidity.

Case study: Summer grafting of *Daphne, Hamamelis,* and *Viburnum*

Summer grafting can be used for a wide range of deciduous shrubs and trees. The choice often comes down to the grower but there may be some plants that do better in the summer, like, *Corylus avellana* 'Contorta' (contorted hazel).

The rootstocks will be potted in the spring and grown on for a season to produce a girth of 6–8 mm. This will often be in P9 size of containers, although sometimes larger containers are used. For example, two-year-old seedlings of *Hamamelis virginiana* rootstocks of 6–8 mm girth are grown in 1-litre-deep containers. Some growers keep the rootstocks well watered and find that the natural reduction in sap flow in late autumn is not a

Rootstock of beech (*Fagus sylvatica*) for side wedge graft in late summer. Although growth is slowing at this time of year, the medium is dried off so it is just moist enough to reduce sap flow. It is interesting to note that other growers find that they can only successfully graft beech in late winter.

Rootstock cut for a side wedge graft being prepared.

Completed side wedge graft of beech.

Summer grafts placed on their side and covered with one layer of contact polythene and then another polythene sheet as a low tunnel. Reducing the air space around the grafts helps to prevent fluctuations in humidity. Grafts are not waxed.

Whip grafts can also be used in late summer if the sap flow is controlled correctly. Note the reduction in leaf size to reduce transpiration.

problem even when using an apical graft. Other growers who use an apical graft find they get better results by drying back the rootstocks prior to grafting and then just water sparingly to control sap flow. The side graft is also used at this time by other growers to act as a sap draw while the graft union is forming. This again illustrates the choices available to the grafter to achieve a successful result.

Scion material should be freshly cut, usually on the morning of grafting, placed in polythene bags and kept in a cool place. In practice, nurseries producing a large number of grafts will not find this feasible and so scion material can be kept in a cold store for several days if required. The scion material is collected from the present year's wood and has mature buds; the wood is firm, and soft flexible shoots or tips are discarded.

The final preparation of the rootstock is to cut to height just prior to grafting. The scions are cut to length and two leaves are usually left on the scion. These can be reduced in size if they are large. The grafts are prepared, tied and may be waxed over the tie if using rubber strips. This is not essential, however, since the grafts are kept at a high humidity while the union forms. The grafted plants can be put in crates on the floor of a tunnel, stood up, or even put on their side on benches. It all depends on the facilities and space available. What is important is the well-sealed, contact polythene around the grafts. Milky polythene (150 gauge) can be used to provide one layer of shade to the grafts. Alternatively, clear polythene can be used as long as it is possible to shade above this on bright days. White horticultural fleece can be used between the grafts and polythene to help reduce condensation droplets falling on the grafts that can cause problems of rot. Whichever type of cover is used, it is important to have moveable shade material to put on, or take off, the grafts when it is a sunny or dull day.

The grafts are kept well sealed until the callus has developed around the union and has started to colour, going from white to green or brown. It is best not to check the grafts too frequently but keep them well sealed. After five or six weeks, weaning can begin by taking the polythene off the grafts at night initially, then altogether after eight to ten weeks. The grafts are then left over winter and watered sparingly initially, in the spring, when they break into growth.

Top-worked grafting

Top-worked grafting to produce small trees or weeping forms is not new and was common in Victorian public gardens. This tree form became popular again in the 1990s, partly to meet a demand for patio plants that would give some height in planters, and also to provide trees suitable for small gardens.

In this application of bench grafting, the goal is to produce quality plants with straight, thick stems and large, even heads in large enough containers to support the weight at the top.

To meet these demands, some growers use two scions grafted onto a rootstock. One scion, however, is usually adequate to produce the required head. It is important to ensure the scion is straight when grafted and the subsequent growth is pruned to produce a well-branched head. The production time from grafting to sale may also be increased from two to three years, enabling a larger plant to be produced and time to prune the head into a fuller shape.

The rootstocks for top-worked grafting can be either bareroot or container-grown. Plants of *Cotoneaster*, *Euonymus*, *Larix*, *Prunus*, and *Salix* are grafted bare root and then potted up, whereas other species like *Robinia* are best grafted from container-grown rootstock. Once again, the choice will come down to the grower's preference and the type of rootstock required.

Case study: Top-worked *Salix caprea* 'Pendula'

The Kilmarnock willow (*Salix caprea* 'Pendula') has been a favourite small weeping tree for many years, although its popularity does seem to be on the decline. There are a few rootstocks that can be used. *Salix ×smithiana* and *S. viminalis* are commonly used and *S. discolor*, the American willow, can also be used. *Salix viminalis* 'Bowles Hybrid' may be the best rootstock as it is a vigorous and hardy variety that has a low amount of suckering and is little affected by aphid.

Whether rooted or unrooted rootstocks are to be produced, parent plants of *Salix viminalis* 'Bowles Hybrid' should be pruned hard back in early spring prior to growth starting. Plants

naturally have a root-to-shoot ratio of 1:1. The harder a plant is pruned, the more vigorous the subsequent growth until the plant can restore the 1:1 ratio. In the process, the plant produces vigorous stems up to 3 metres long.

It is also possible to produce Kilmarnock willow from cutting grafts. Instead of rooting the rootstock one year and grafting the next, it can be grafted and rooted at the same time. This is possible since willow is an easy-to-root species that will quickly grow and enable the scion to reach a good size in one season.

To produce traditional rooted rootstocks, leafless winter cuttings 15 cm long are prepared and struck into the soil. A sheltered site should be

Stock plants of willow rootstocks. Hard pruning in the spring produces vigorous growth that can be cut as long stems for cutting grafts, or prepared as conventional cuttings to get a larger number of rooted rootstocks after a further year.

chosen and the soil should be free draining and weed free. The soil is prepared by cultivating to form a fine tilth, and black polythene is laid over the soil and dug in. This will act as mulch that will help to warm the soil in the spring and promote rooting, and will also suppress weeds. In addition, lines of irrigation pipe, leaky pipe for example, can be put down below the polythene so the cuttings can be kept well irrigated. The prepared cuttings are pushed through the polythene so that two thirds of their length is inserted. Cuttings stuck in early spring will have grown up to 3 m in height by late summer. After leaf fall in autumn, these stems can be dug up, graded to 120 cm or 80 cm stems, with a girth of 12–14 mm and

14–20 mm and stored until they are required for grafting.

Cutting grafts have the advantage that the season spent rooting the cuttings can be omitted. The stems originally collected from the parent plant are graded to the 120 and 80 cm sizes, with girths as above, instead of being made into 15 cm cuttings, and stored until they are used for grafting. The disadvantage of this system is that a greater area of stock plants is required to get the number of stems required of the correct grade for grafting.

Once the rootstock is selected, it is time to select one or two scions. If a single scion is to be grafted to the top of the rootstock, then it is important to select straight scion shoots, as a bent

Leafless winter cuttings inserted through black polythene. The polythene absorbs heat from the sun and warms the soil below to encourage rooting. It also acts as a mulch, suppressing weed growth and reducing moisture loss from the soil.

Unrooted willow rootstocks and prepared scion sticks prior to grafting.

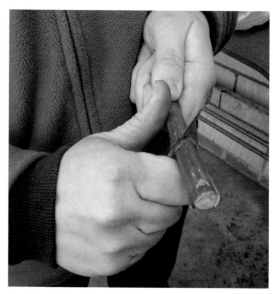

Two scions are going to be put on this rootstock. A cleft graft is prepared below the apical wedge graft on the opposite side of the rootstock.

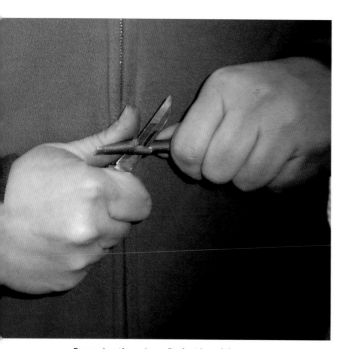

Preparing the scions. Both sides of the scions are cut, with a slightly shorter cut being made for the cleft graft.

The scion is set into the rootstock cuts at a slight angle to ensure four good points of cambial contact.

Final graft with the scion ready to be tied in to the cleft graft.

scion will produce a one-sided head. A single scion will produce a good size of head if pruned early on to encourage branching. In the case of the quick-growing willow, this will make a good, saleable plant in one season.

Sometimes it is preferable to graft two scions on the rootstock to produce a larger head in one season. Although this is probably more important with slower growing species, it does avoid problems of an unbalanced plant if a scion is bent. Again, the grafts used come down to personal preference. An apical wedge graft can be applied at the top of the rootstock. The scion can be attached at a slight angle to ensure that there are four points of contact for the rootstock and scion cambium. This avoids problems of poor cambial contact if the cuts do not precisely match. This technique can be

A chisel graft can be used where the scion has a much smaller diameter than the rootstock. This can also be used for changing varieties on mature fruit trees by top working.

Scion and rootstock being fitted together. A chisel graft gives cambial contact on two sides of the scion and is preferable to using a whip graft, which applies the scion to only one side of the rootstock.

Polythene bags placed over top-worked grafts to maintain the humidity around the graft union.

A grafting machine used for preparing saddle grafts. Good for larger diameter grafts and harder wood.

Tying-in the graft. The union has to be tied tightly to pull the cuts together and form a good union.

used, as well, with modified side veneer grafts. This graft is tied in with large 240 mm by 8 mm ties.

The second graft is carried out on the opposite side of the rootstock to the first. A side cleft graft is used which has a short cut at the rootstock. This angles the scion slightly away from the rootstock and allows the graft to be made higher up the rootstock to keep a compact head. The second graft is tied in with a smaller 200 mm by 6 mm rubber band, and then both grafts are dipped in wax. It may be beneficial to rotate the graft while waxing to avoid the formation of air bubbles within the wax.

If cutting grafts are used, then the base of the rootstock is soaked in a bucket of water for about a few days or more prior to grafting. After grafting, the plant is inserted into a container with growing medium (so that rooting will be initiated) and placed in a greenhouse. As long as you push the right end of the stem into the medium, it should root well. If using rooted rootstocks, then the plant is potted up after grafting in the usual way and placed in the greenhouse.

An alternative to waxing is to place polythene bags over the graft unions in the top-worked plants and tie below the lowest cut. This creates a localized area of high humidity that prevents the graft from losing water.

Conifers

Because most conifers are evergreen plants, they are not suitable for budding. Rather, bench grafting is used. Unlike broadleaved plants, conifers do not seem to have an option but to use a side veneer graft, or one of its variants, for a successful union to form. This type of graft, where the top of the rootstock remains on the plant until the graft has taken, is used for two reasons. Firstly, the rootstock has to be in active growth when the graft is prepared. Retaining the top growth of the

rootstock acts as a sap draw so that the pressure of water rising through the stem by evapotranspiration will not push the scion off the rootstock before the union has formed. Secondly, it is often used where the graft union is slow to form. The top of the rootstock then provides nutrients and photosynthates to nurse the scion through the healing process. At optimum temperatures, the cambial contact may take 21 to 42 days in *Picea abies* and *Pinus sylvestris*, up to 50 to 90 days in *Picea sitchensis*, and only 7 to 21 days in *Malus* and *Pyrus*.

For conifers, the optimum is a base temperature of 18–20°C and an air temperature of 16–18°C to promote callus formation while keeping the scion buds dormant. This can be achieved by using a heated greenhouse of no more than 18°C, or providing base heat under the grafted plants of 18–20°C while maintaining an air temperature above the grafts of less than 18°C.

Graft failure in a number of conifer species has been due to low water potential in the scion. The lack of turgor prevents cell expansion and therefore callus, the initial healing stage, does not form. It is therefore important to maintain a humidity of 80 percent to prevent excessive transpiration through the leaves of the scion. Traditionally, the correct conditions are provided by an open bench with mist irrigation or a deep, closed case covered with glass. In both examples, the potted rootstocks are placed into peat so that the graft is covered to ensure moisture is retained around the union. During the time the graft remains in the case, the peat must be maintained at the correct moisture. Squeezing a handful of the peat assesses this. A few drops of water should come through the fingers. If water flows easily, the peat is too wet, but if no drops can be produced, then it is too dry.

The grafts are inserted at a 45-degree angle with the scion facing upwards. This reduces the depth of peat required to cover the grafts, and the scion at the top reduces the risk of water running into the union. Syringing can be used to add small quantities of water that may be required and the glass has to be removed and drained. All this is necessary so that irrigation water or condensation will not get into the graft union and cause it to fail. This is a very time-consuming process that requires a very skilled person to manage.

In the late 1970s, a polythene tent system was developed for the aftercare. This produces a higher and more consistent humidity than using glass or mist. The polythene is laid over a curved frame that is tall enough to leave an air space above the grafts. This ensures that water droplets will not drip on the grafts. Polythene with an anti-condensation treatment is now used to prevent further drips on the graft. The anti-condensation treatment allowed the polythene to be laid in contact with the grafts, but this was found to increase the problems with *Botrytis* that can arise in the warm humid conditions. This tent system means that the grafts no longer need to be covered by peat (although they may be stood on peat to help maintain the correct humidity). The plants can be stood up, allowing more plants to be placed under the grafting tent, and the aftercare becomes much more straightforward. This has reduced the time and effort it takes to care for the grafts and it can give a greater percentage success than the traditional method.

Rootstock preparation

Bareroot plants are not suitable for side veneer grafts as the rootstock must be in active growth prior to grafting. 1+1 transplants or 1u1 undercut open-ground seed-raised plants may be bought in as rootstocks, potted into P9 containers or similar, and grown on for a season to ensure a well-established root system. The other requirement is that they have at least a 6 mm diameter stem near the base of the stem. It is important that this matches

the diameter of the scion, so if larger diameter scions are produced by more vigorous species, then larger diameter rootstocks will be required. Many propagators are now purchasing 1-year cell-grown plants that have been grown under protection prior to being hardened off in late summer. This rapid early growth is thought to give a better, straighter stem compared with field-grown plants that can be more knotty at the base.

These cell plants are usually potted into P9 containers and grown on for a season prior to grafting. However, it is possible to graft directly onto the rootstocks at the cell stage if they have a suitable stem diameter. It may require working closely with your cell plant supplier to obtain cell-grown plants that have the required stem diameter and can be used for grafting in the first year. In practice, some plants are grown on in P9s if they do not have the correct diameter. This will cause a few difficulties in managing different sizes of rootstocks, and further work with the rootstock producer to have more precise grading, may be worthwhile. The advantage is that the time and cost of growing on the rootstocks is avoided and a smaller space is required for the aftercare of the grafts.

The actual species bought in as rootstocks will depend on the scion species and cultivars to be grown. Possible conifer scion and rootstock combinations are given in the appendix. Where there is a choice of possible rootstocks to use, the choice may depend on what is available, or what soil conditions the plant will be grown in eventually. It may also depend on the quality of graft that can be obtained. For example, *Pinus sylvestris* is a common rootstock for two-needle pines but it does not give the best root system and is prone to root rots. *Pinus mugo* gives a better, bushy dense root system but it has very branched and knotty angular growth; finding a straight stem to graft onto can be difficult. *Pinus uncinata*, which is closely related to *P. mugo* and has a similar root system, produces a straighter stem and is easier to graft onto and is now preferred by some growers.

The rootstocks are started into growth by bringing them into a greenhouse two or three weeks prior to grafting and keeping them at a temperature of 10–12°C. The rootstocks will have been kept outdoors over winter to receive the correct cold requirements. Sap flow will commence in the warmth of the greenhouse and this is controlled by drying the medium to about half the amount of moisture it had when kept outdoors. Sap flow tends to be greater in the P9 containers than in cells. The quality of the rootstock is the key to successful grafting and there needs to be a few millimetres of fresh white root visible on the rootstock prior to grafting.

Rootstock preparation depends on the species. For *Pinus*, the top buds are left but the sides are

Rootstock of spruce ready for grafting.

Rootstock of pine prepared ready for grafting.

trimmed. With *Picea*, the top is cut down and the side shoots reduced in length but not removed entirely. The overall rootstock height will be up to 30 cm at grafting. Just prior to grafting, the stem is cleaned where the graft will be carried out, to avoid the top of the scion coming into contact with the rootstock foliage which could cause disease problems. Trimming the lower shoots of the rootstock will create a clear space for the scion. In addition, the scion should be attached on the outer bend of the rootstock. It is unlikely to be perfectly straight, to keep the scion clear. The scion should be attached as low down and near the crown as possible.

Scion preparation

The scion material is collected no sooner than 10–14 days before grafting and should be collected when the parent plants are not frozen. It is preferable to graft soon after collection and the same day if possible. One-year-old shoots are collected for the scion about 10–15 cm long. They should have a terminal bud and at least three radial buds. If the shoots are not thick enough to match the rootstock diameter, then two-year-old material may be used. Prevent water loss from the scions by placing them in a polythene bag or wrapping them in moist hessian to keep them cool.

In the United Kingdom, conifer grafting is usually started in late winter. The order of grafting conifer species is often *Picea*, *Pinus*, *Abies*, and then *Cedrus*. Although many growers start with *Abies* and *Cedrus* (as these species break bud first), leaving them to last reduces their time under polythene as they are prone to *Botrytis*.

The scions are prepared immediately prior to grafting by trimming the needles from the base up the stem of the scion-wood to about half the length of the scion. With pines, the cones are removed if present. Either the side veneer graft or side wedge graft can be prepared. Some grafters

Stock plants for the scion graft wood. Each scion needs a healthy terminal bud and secondary buds that will give a well-branched plant.

The cut should not be too deep and should not go into the pith.

Completed grafts side veneer grafts waiting to be put into trays. Note the white growth visible on the rootstocks. This is mycorrhiza—beneficial microorganisms that aid the uptake of nutrients through the roots.

In a side veneer graft, the second cut is made above the first on the stem.

The cut must be straight down the rootstock, not scooped.

In a side veneer graft, a close fit between scion and rootstock must be achieved.

The rubber tie is overlapped but not waxed before the graft is put under the polythene tunnel.

Twenty grafts are placed in each tray. Only half the height of the cell is put into the medium to reduce the weight. All the unions face the same direction to make heading back easier.

Grafts under the polythene tent to maintain a high humidity.

prefer the side wedge as it gives more cambial contact than the side veneer, which can be more easily damaged immediately after the union has formed. Nonetheless, the side veneer is probably easier and quicker to prepare, requiring fewer cuts. It is important not to cut into the pith of the scion. That is, do not slice too deeply into the wood. It is easier to do this if cutting only one side in the veneer graft. At the end of the day, either technique will produce plants of excellent quality.

The grafts are tied in using 140 mm by 4 mm rubber bands and again techniques vary slightly as long as they achieve the requirement of secure ties applying pressure at the graft union. The grafts may be wrapped 5–6 times, leaving some of the union exposed between turns. The graft is then waxed to prevent moisture loss, which may occur even under the polythene. The wax will also ensure any drops of water do not enter the graft union. Alternatively, the ties can be overlapped, covering the whole cut surface and giving as good a result as waxing, although the tying takes longer to complete. Depending on the method used, a rate of between 40 and 60 grafts an hour can be achieved by experienced grafters.

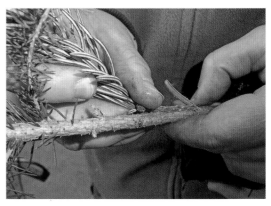

The scion is inserted onto the rootstock and the flap covers the cut at the back of the scion.

The graft is waxed so the rubber tie is not overlapped.

In a modified side veneer (wedge) graft, a single cut is made down the rootstock. The bark is not removed.

The wax is applied by brush, as the graft cannot be dipped since there are leaves on the scion (and rootstock).

However the graft is prepared, care is required at watering not to wet the foliage. Not only may this have a detrimental effect at the graft union but it may also encourage *Botrytis*, which is a major disease problem during the formation of the graft union.

Aftercare

Aftercare of grafts is just as important as the quality of the rootstock and the grafter's skill. Good aftercare has five main aims:

1. To maintain temperatures that are warm enough to keep the rootstock growing and the callus to form but do not get too hot.

2. To provide high humidity and thus prevent desiccation of the scion prior to the union forming.

3. To limit irrigation initially to control sap flow from the roots and to prevent water entering the union by avoiding overhead irrigation until the union has formed.

4. To reduce the top growth of the rootstock so that it supports the scion for long enough but does not overcompete for photosynthates and nutrients.

5. To prevent disease problems, especially *Botrytis*, by ventilation and heading back the rootstock to allow airflow around the scion.

There are a number of ways the correct aftercare can be achieved and the examples given are from two nurseries that produce quality plants. What they have in common is their attention to detail and a thorough understanding of their system so that they have a high grafting success rate.

Commonly, grafted conifers are grown in P9 containers and the graft union is waxed. These are then placed in a glasshouse kept under 15°C. Although this temperature is lower than optimum

temperature, because the graft union is waxed, it is possible to keep the grafts on an open bench to reduce humidity and the risk of *Botrytis*. Only as light levels increase does the nursery put fleece over the grafts to retain a higher humidity.

Once the grafted plants show callus development or the scion starts to grow, then the plants are headed back in three stages.

1. Callus is normally visible after six weeks and rootstock shoots on the side of the scion are cut back to avoid contact between the shoots of the rootstock and scion. If there are any signs of *Botrytis* this is cut out first.

2. After about 10 weeks, the rootstock growth is cut back to three shoots. This keeps the rootstock growing and supporting the scion.

3. After about six months the plant is potted on, into a larger container and at this point the rootstock is cut right back.

Watering and venting are the other important requirements for the aftercare of these plants. Watering should be limited to keeping the medium moist initially after grafting. Overhead watering is to be avoided. Once the graft has taken and the scion is growing away, watering can be done overhead as required. At first, the plants should be vented every 10 days, increasing to twice a week once there are signs of growth. After the first heading back, venting should be increased to every day.

Alternatively, a slightly different technique is to use cell-grown rootstocks. The grafted plants are plunged into peat in 40 cm by 60 cm trays. Only half the depth of the cell is put into the peat to ensure that the graft union is not covered and to reduce the weight of peat in the trays. Twenty grafts are put in each tray and the unions are all pointed in the same direction to make heading back easier. The trays are placed under a

The first reduction of the rootstock removes side shoots, especially around the scion.

Shortening the top of the rootstock to divert more sap to the scion growth.

Final stage of heading back the rootstock.

polythene low tunnel within a walk-in tunnel. The walk-in tunnel is double skinned with bubble wrap between the skins. This gives 5°C frost protection and traps daytime heat.

The first grafts in late winter are kept covered for ten weeks. Since the cover is removed from all the grafts at the same time, the later grafts are only covered for five to six weeks. The grafts are vented once a week when they are checked for water. The grafts are not watered overhead. Only the peat is watered as this keeps the foliage dry.

Through early spring, watering is carried out sparingly and may only be required two or three times. Venting of the tunnels is increased in midspring: the sheets that used to be lifted up on hot days and closed again at night are now removed entirely. The walk-in tunnel is kept closed initially to prevent too large a drop in humidity and is vented as required by opening the end doors. It is planned to put a fan into the house to improve the air circulation further. By late spring, the grafts have sealed and the plants can be watered overhead. There is a particular demand on water from *Picea* and *Pinus* species in early summer, when soft, supple growth is developing. As this new growth starts to harden, the water requirement reduces.

In this example, to reduce the risk of disease, the grafts are headed back in one operation in late spring to early summer once the scion has started to grow. This is a selective pruning depending on the scion growth of each graft. If there is not enough scion growth, then heading back may be delayed to midsummer. The selective pruning is made easier by having all the graft unions facing the same way in the trays. Heading back in one operation may not be as good a technique as heading back in stages, but has proved successful for the system described above.

The grafted plants are potted on in late summer. The roots have grown from the cells into the peat in the tray. Extension growth has more or less finished and the new growth is firming by this time. Therefore, although there is some root damage when the cells are removed from the trays, there is not too much stress caused to the plants, as they are not in active growth.

Whip and whip-and-tongue grafts

The whip graft is a good starting point when learning how to bench graft. Once you can carry out this cut, then you can adapt the knife skills to other grafts.

The equipment required for this graft includes secateurs, a grafting knife, grafting tape, cleaning materials (alcoholic wipes or methylated spirit for disinfecting the knife and secateurs between cuts), a sharpening stone or leather strop, labels, and pencil or indelible marker. Set up the workstation so that everything for grafting is at hand and you do not need to make unnecessary movements to locate tools and supplies.

The secateurs should have a sharp blade and tight action (good tension) so that they make a clean cut. Do not use secateurs to cut nylon string or open polythene bags, as this will damage the edge of the blade (the same goes for grafting knives). Do not cut wood thicker than 20 mm as this will cause repetitive strain problems in the hand and can loosen the tension of the secateurs so they make poor cuts. It is worth investing in a pair of quality secateurs with replaceable blades, high-quality steel, easily adjustable tension, and easy to dismantle and clean. If just using for propagation purposes, a lightweight pair is also ideal.

The grafting knife should have a high-quality, razor-sharp blade. It should be sharpened on one side of the blade edge only. (Note: most knives are sharpened for right-handed use. Few manufacturers make left-handed grafting knives but they are available. If you only have a right-handed knife,

a technique for left-handers is described later). Grafting knives come in two styles: a straight blade, or a broader blade that is straight and then curves up at the end. Some people prefer the latter knife as they find it finishes off the cut through hard wood more easily.

Three types of tape are commonly used in grafting. Polythene tape 26 mm wide both holds the graft and seals it. The exposed cut top of the scion can be sealed with petroleum jelly to prevent desiccation. Polythene tape does not break down in sunlight and so has to be removed by the propagator.

Rubber ties can be used instead of polythene tape to tie-in a graft. Various sizes of bands are used depending on the diameter of the rootstock. For *Acer palmatum* and other rootstocks 6–8 mm in diameter, strips are typically 140 mm long by 4 mm wide. For Rosaceae species and other rootstocks 8–10 mm in diameter, the strips are usually 200 mm long by 6 mm wide. For larger

Styles of grafting knife used for bench grafting. Top: Straight bladed, sharpened on one side only, specifically designed for bench grafting. Middle: Larger blade with curve at end. Traditionally used for field grafting but some prefer it for bench grafting. The curve helps to finish off the cut. Lower: Disposable, craft type blade; these knives do not require sharpening and the blades are changed regularly. The blade must be strong enough and straight to use for grafting. It overcomes the problem of sharpening knives to the correct sharpness.

diameters and for top-working where two grafts are attached, rubber ties are usually 240 mm long by 8 mm wide. Unlike polythene tape, rubber ties must be sealed. To do this, grafting wax is used. The container for the graft must be deep enough so that the wax will cover the scion and all the cut surfaces when the graft is dipped. Special equipment for melting the wax is available that will maintain the optimum temperature of 60–70°C. If this is not available, then a container of wax can be heated in a larger container of water on a hot plate. The rubber is biodegradable and so does not need to be untied.

A third option for tying-in grafts is paraffin (Buddy) tape, which holds and seals the graft. This tape is self-adhesive and does not require a knot when tying the graft. The tape breaks down in sunlight but only if it is not wrapped too much. Only wrap 2–3 times, half a patch is usually adequate, and it should degrade. If wrapped more, then it may not degrade and is very difficult to remove. The exposed cut top of the scion can be sealed with wound sealant or petroleum jelly to prevent desiccation. This type of tie is not suitable for harder wood that requires the scion to be tied tightly to the rootstock.

Collecting and preparing the scion

For winter grafting, the bud-wood should be collected in midwinter and stored below 5°C to ensure the buds are fully dormant when they are grafted. In the summer, the bud-wood should be collected as near to the time of grafting as possible, kept cool, and sprayed with water regularly.

The selected scion-wood should be firm, not too soft and pliable, but also not too stiff and woody. It should be as near the diameter of the rootstock as possible, avoiding thin, weak growth. In most cases, the scion should consist of the present season's growth containing healthy buds, but for summer grafting of beech (*Fagus*) and similar trees, it

may be beneficial to have a portion of second-year wood at the base—especially if the wood is a little thin. Ensure the growth habit of the scion-wood on the stock plant is of the correct character. If, for example, collecting from a weeping form, do not collect horizontal- or upright-growing shoots. Avoid any diseased wood and prune this from the stock plant at a later date.

The length of scion and number of buds will depend on the species being grafted. Usually three buds are sufficient as long as the length of the scion makes it easy to hold while cutting. If a single leader is to be grown once the graft has taken, then only one bud is left to grow once the graft has taken. It may, therefore, be feasible to use fewer buds, especially where there are long internodes. If preparing an interstock that will be cut top and bottom, then a longer shoot is required, about the length of secateurs handles.

The standard scion preparation is to make a sloping cut about 5 mm above and away from the top bud. Then make a straight cut at least 20 mm below the lowest selected bud. This will enable the bottom bud to remain behind the cut after it has been made. As discussed earlier, this is not essential although it is still preferred by many grafters who believe it makes a better union. Others cut the scion to length and are not concerned about having a bud at the back of the cut.

Where internodes are long, or if a hot-pipe callus system is used, it may be better to cut the scion to length and not have a bud at the back of the cut. The scions may be graded to size (no more than three size grades) and they must be labelled and placed in a box ready for use.

Preparing the rootstock

The stems of the rootstock should be cleaned and cut back to the required length—20 cm for fruit trees and 12 cm or even lower for ornamental trees and shrubs. If the grafting is being used purely as a method of propagation, then cut as low down as possible, but at least 40 mm long for the graft cut. Some propagators prefer to cut the rootstock just above a bud, as they believe it reduces the risk of any dieback occurring on the rootstock. This is not essential but a small "step" should remain at the top of the rootstock after preparing the cut. The step will reduce the risk of any dieback.

For rootstocks from bareroot plants, ensure that fibrous roots are present and that they are healthy and not dried out. Prune the roots to fit the size of container in which they will be potted. Some species, like quince, have roots that tend to dry out quickly. The roots should be soaked in water after pruning and then potted immediately after grafting.

Container-grown rootstocks will have been growing in a container for a season prior to grafting. The size of the container depends on the species but normally is around the P9 size. The rootstocks can be grown outdoors for a season prior to grafting and brought in a few weeks before grafting to dry off and start root growth. Alternatively, they can be grown under protection and watered sparingly over winter and as root growth begins to control sap flow.

Making the cuts for right-handers

The instructions given here are for a right-handed person. In cutting the scion, the aim is to make a straight cut 40 mm long that runs from one side of the bud-stick to the other. The key to success is to pull the knife and wood straight apart without moving either your fingers or wrist. The knife should slice through the wood starting at the heel of the blade and finishing the cut at the toe. Because the knife is sharpened on only one side, the back of the blade is flat. This means that it will move straight through the wood at whatever angle you position it, at the start of the cut. To get the 40 mm cut, imagine a straight line running along the

back of your knife, entering the wood where the blade is positioned and exiting at the other side 40 mm further on. It will in fact be almost flat on the scion-wood. Remember that grafting involves quite small movements close to the body, not large and expansive movements.

With your hand like a claw, hold the bud-stick in the left hand, leaving the base of the scion protruding by 50 mm (Figure 5-2). If the scion is curved, cut on the back of the curve, as this is easier to get it straight than if cutting on the curve. If you are keeping a bud behind the cut surface, then have the protruding bud facing you.

The knife must be held at an angle almost horizontal with the ground, **not** with the blade vertical, so that it will slice through the wood. Hold the knife across the palm of your hand with the edge of the blade towards your thumb (Figures 5-3 and 5-4). Hold the knife in your fingers, rather than your fist, and wrap your index finger around the back of the blade. Try to use a relaxed grip to reduce tiredness and the risk of repetitive strain, and keep your thumb away from the edge of the blade.

Bring your hands together in front of your chest (Figure 5-5 and 5-6). The scion should be parallel to your chest with the bottom bud pointing

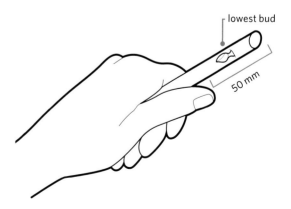

Figure 5-2. Position of hand and bud-stick prior to grafting.

directly at your body. The knife and thumb pass either side of the scion with the knife on the back of the bud-stick. The scion should be held between the ball of the thumb and the heel of the knife blade. Do **not** press your thumb on the scion. The thumb will act as a guide only.

Your hands should be close in to your body with your elbows out to the side. This is sometimes called a butterfly cut as your arms look like the wings of a butterfly. By having your hands at chest height, they will move straight apart. If you hold them lower down, then your hands move in a curving action and you do not get a straight cut.

Begin the cut about 40 mm from the end of the scion and gently draw the blade along. Your elbows should move straight out and up (**not** backwards or downwards). Concentrate on moving your weaker arm (the one holding the scion), as your stronger arm will move automatically. Do **not** squeeze with your fingers or bend your

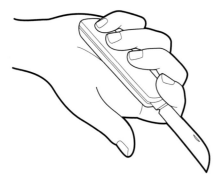

Figure 5-3. The position of the knife across the hand with the edge of the blade towards the thumb. Note that the index finger wraps round the back of the blade, not the handle.

Figure 5-4. The final position of the knife and the stem. The knife should be almost flat against the stem to get the length of cut required. Make a slicing cut from the heel of the blade to the toe.

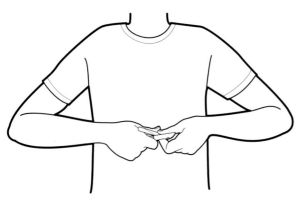

Figure 5-5. Position your hands in front of your chest. This will help your hands to move straight apart. If held lower down you cut in a curve. Tuck your hands into your body and push your elbows upwards and apart to make the cut. It should be one continuous movement but not too fast.

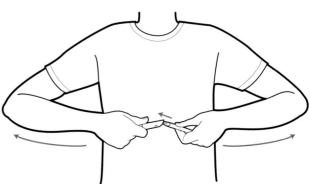

Figure 5-6. The grafting cut moves your hands about 40 mm apart. At the end, your elbows should still be bent and you must not squeeze your hand or bend your elbow, as this will prevent a straight cut.

wrist; the knife will cut the correct length if set up at the correct angle (the flatter the knife is to the scion-wood, the longer the cut). Restrict and control your arm movements; remember the cut is only 40 mm long! Ensure the knife moves with a slicing action from heel to toe and use the full length of the blade. This completes the cut for a whip graft.

To prep a scion for a whip-and-tongue graft, you will want to add a cross cut (tongue) to the cut made for a whip graft. To add a tongue to the graft, turn the knife and stem so they are pointing away from your body and your hands are together (Figure 5-7). Place the knife one-third from the top of the cut for the rootstock and one-third from the top for the scion. Turn the knife with your wrist to slice down the length of the wood for about 10–15 mm. Take special care with this cut.

The rootstock for both whip and whip-and-tongue grafts is cut in the same way as the scion if it is bareroot material. If it has been cut just above a node, then have the bud facing you and make a 40 mm long cut. The depth of the cut will depend on the breadth of the cut on the scion. If the scion is thin, then a shallow cut can be made on the rootstock so that both sides of the cuts match. Even if the scion and rootstock diameter are closely matched—remember to leave a small step on the rootstock.

If the scion and/or the rootstock are thicker, or if the rootstock has been pot-grown, then a different method of making the cuts can be used. In this method, the cut is made towards the body. The scion is held with palm of the hand facing upwards and the bottom bud protruding from below the little finger. The elbow is held against the body so that the forearm and scion are almost at right angles to the body. This keeps the hand steady. The lowest bud faces towards your arm. The knife is turned so the edge of the blade is vertical and the point is aimed away from you. Position the

knife on the scion as before and pull the knife through the wood in a slicing action (Figure 5-8). For the pot-grown rootstock, hold the pot on a bench facing towards you at an angle and cut with the knife as above.

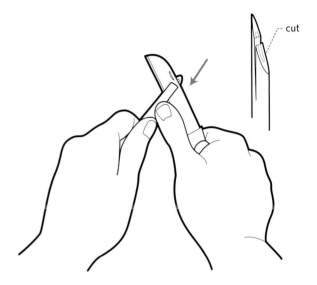

Figure 5-7. Position of hands at start of cross cut

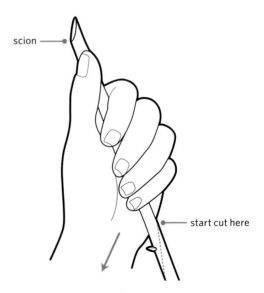

Figure 5-8. Hold the stem facing towards you and position the knife on the stem as before. Pull straight through the wood in a slicing action.

Making the cuts for left-handers

For left-handed people, obtain a left-handed knife and proceed as above but in mirror image. If using a right-handed knife, hold the scion in the right hand facing vertically towards the floor (Figures 5-9 and 5-10). Place the knife angled on the scion with the knife facing as near upright as possible. Your elbow should be bent and the hand turned so your thumb is pointing downwards. The knife is held in the hand as for the right-handed grafter but with the edge of the blade facing away from your hand. The knife is held vertically with elbow bent. Place the heel of the blade on the wood, 40 mm from the end and almost flat so it will give the correct length of cut. Pull your hands straight apart.

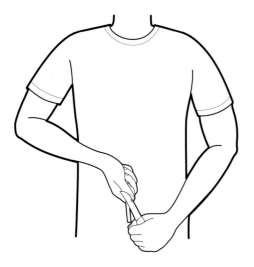

Figure 5-9. Grafting left handed with a right-handed knife can be done. The knife and stem are held almost vertically with elbows bent.

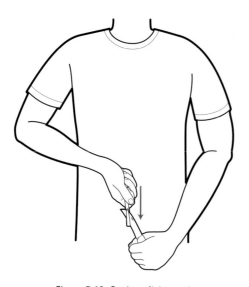

Figure 5-10. Cut in a slicing action.

Troubleshooting

If the cut ends up being scooped instead of straight, the usual cause is a forward wrist movement associated with a tightening of the grip.

Often the cut leaves a small, thin tail at its end. To prevent this, apply a little more pressure with the thumb behind the wood being cut, but take care not to push the thumb on the edge of the blade. Using a grafting knife with a curved point may also reduce this problem.

When the cut is made horizontal with the chest, a tail may be caused due to the elbows moving backwards or downwards or the non-cutting arm straightens. Remember: the arms must remain bent and the elbows move straight out and up.

When the cut is made towards the body, a tail may be caused by movement in the non-cutting hand. This problem can also be a result of forcing the knife through rather than allowing it to slice along the full blade length. Remember not to hold the knife vertically. Instead, hold it almost horizontal and slice through the wood from heel to toe.

Sometimes the wood may become twisted along the length of the cut. Twisting results from a rotation of the arm usually at the wrist.

Aligning the scion and rootstock

Hold the cut surfaces together using the thumb and forefinger. If a tongue has been added, then the tongues are pushed together and this will hold the graft together prior to tying. In either case, do **not** handle the cut surfaces as dirt or grease from your fingers can cause a barrier to the graft forming.

A small section of the scion cut must remain visible above the level of the rootstock (often called a church window) to ensure that the cambium of the scion and rootstock match up from the top of the union (Figure 5-11). The width of the cuts should match as closely as possible, but ensure that the scion is not wider than the rootstock. It is suitable to have a slight edge of cut rootstock visible behind the scion. If the scion is very much smaller than the rootstock, then it can be put to one side so that one section of cambium matches up. Do not put a small scion in the middle of the rootstock cut as the gap between the cambium will be too great and the union will not form. If only thin scion is available, then take a smaller slice out of the rootstock when making the cut, or

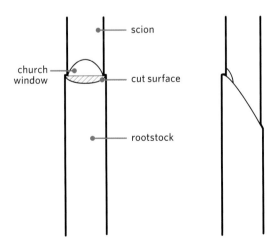

Figure 5-11. When joining the rootstock and scion, align the cambiums together and leave a church window to ensure that the cambiums match from the start.

use a different grafting technique—for example, a rind graft. Once the cambiums are matched, then the graft can be tied.

Tying-in and waxing the graft

Normally the tie will be started at the top of the graft, on the scion just at the top of the rootstock, although it can also be tied from the bottom up. Hold the end of the tie firmly with the forefinger and turn the tie round the graft, moving upwards to cover the top of the cut and covering the end of the tie to hold this in place. Stretch the tie slightly to ensure the correct pressure is applied. Angle the scion towards the body and wrap the tie around the scion and rootstock, maintaining the tension all the way down.

If using polythene or paraffin ties, then ensure the edges overlap but only slightly. If using rubber ties, overlap the edges or leave a small gap between turns as it will be waxed. Go beyond the bottom cut by a turn. Unless the tie is self adhesive, make a loop using the forefinger of the hand holding the graft and pass the end of the tie down though the loop with the other hand to secure. Ensure the tie keeps going in the same direction so that the tension is not released.

An alternative is to start in the middle and wrap down below the union, then up above the top of the union, and then down to tie off in the middle. This may be an easier method when tying a whip graft, which can be awkward to hold and tie, but it is slower and uses more tape then just tying from the top down.

It is easiest to dip the completed graft into a container of molten wax, ensuring the wax is at the correct temperature, 60–70°C, and all the cut surfaces are covered. It may be beneficial to do this with a twisting action to reduce the chance of air bubbles forming under the wax. If the graft is too big to be dipped in a container, then the wax can be painted on with a small brush. If the rootstock is pot-grown, then make a simple cardboard disc the diameter of the pot to protect the growing medium from falling out; cut a slit to the centre and put this cover over the pot before dipping it.

Polythene and paraffin ties do not require waxing, but the top scion cut should not be left exposed and this can be sealed with wound sealant or petroleum jelly.

Side veneer graft

The side veneer graft is used for conifers and deciduous species, adjusting rootstock diameters and scion lengths. The equipment required for this type of graft is identical to that required for whip and whip-and-tongue grafts: secateurs, a grafting knife (either straight bladed or a craft type), grafting tape, cleaning materials (alcoholic wipes or methylated spirit for disinfecting the knife and

secateurs between cuts), a sharpening stone or leather strop, labels, and pencil or indelible marker. Set up the workstation so that everything for grafting is at hand and no unnecessary movements need to be made to locate tools and supplies.

Collect scion-wood about 80–100 mm long and of the same diameter as the rootstock, or as near to this as possible, with healthy buds present. Normally a terminal bud with three or four secondary buds for each scion is sufficient. Select the present season's growth that is neither too pliable nor too stiff and woody.

Hold the top of the scion and trim the needles off half the length of the scion. Make a straight cut down the clear stem towards the base 25–30 mm long. Slightly angle the cut, but do not cut too deeply into the pith. Make a reverse cut at the base about 3 mm long and at an angle of about 20 degrees (Figure 5-12 left).

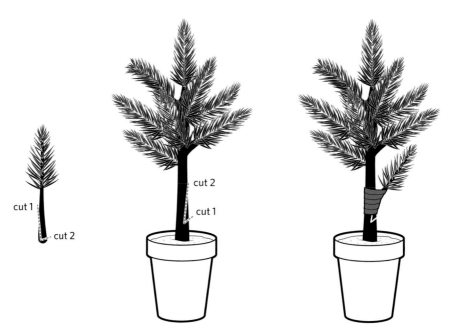

Figure 5-12. Side veneer graft: (*left*) Two cuts are made on the scion. (*centre*) Two cuts are made on the rootstock, both straight and not scooped. (*right*) Scion is attached to rootstock.

To prepare the rootstock, cut it back to 150–200 mm in height and remove needles from the base of the stem. Make a bottom cut at an angle of 20–30 degrees as low down as possible, to a depth of 4 mm (Figure 5-12 centre). A second cut of 25–30 mm is then made down the side of the stem to meet the first cut. This must be a straight cut not scooped, and is done with a slicing action, starting the cut with the toe of the blade and slicing down to the heel.

Attach the scion to the rootstock cut and hold in place with a rubber tie (Figure 5-12 right). The tie does not have to overlap as the graft will usually be put under a polythene tent, but overlapping can be used, especially if the graft is not being waxed. Wax can be brushed on to seal the graft and to prevent moisture dripping into the top of the union.

Modified side veneer graft

A modified side veneer graft, sometimes known as a side wedge graft, is similar to the side veneer, except that the bark cut on the rootstock is not fully removed and is attached to the back of the scion. Only one cut is made on the rootstock, a vertical cut 25–30 mm down the stem (Figure 5-13 centre).

The first cut on the scion is also the same as with the side veneer, but the scion is then turned 180 degrees and a second cut, about 5 mm shorter than the first one, is made down the back. A third cut is made at the base of the scion at a 30-degree angle, sloping away from the shorter cut side (Figure 5-13 left). The scion is inserted behind the cut bark and tied in as before (Figure 5-13 right).

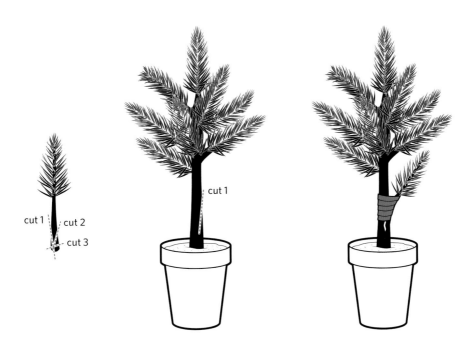

Figure 5-13. Modified side veneer graft: (*left*) Three cuts are made on the scion. (*centre*) A single cut is made on the rootstock. Do not go too deep and do control the knife so the cut bark remains attached to the rootstock. (*right*) Scion is attached to the rootstock.

6 Field Grafting

THE TERM *BUDDING* IS USED FOR ANY GRAFT that uses a single scion bud. It is usually carried out from midsummer to early autumn when the cells of the vascular cambium are actively dividing so that a union is rapidly formed. Two main methods are used: T-budding and chip budding, although T-budding has several variations, such as inverted T-budding or stick budding.

T-budding

Up until the late 1970s, T-budding was almost exclusively the budding technique used for fruit and ornamental trees, as well as roses. T-budding is a shield type of graft, that is, the scion bud is placed behind the bark of the rootstock and adds an extra piece of wood to the rootstock. The advantages of this type of graft are that the bark holds the bud in place and maintains moisture around the bud. Only a rubber patch is needed to hold and seal the graft.

One disadvantage of this type of graft is that there is no direct vascular cambium contact between the scion and rootstock and so the union can be slow to form in some cases. Union canker (*Nectria galligena*) can be a problem as spores can enter behind the lifted bark. There can also be problems of poor bud take because of water entering at the T incision at the top of the bud. To overcome the problem of water, the inverted T-bud is used for susceptible species like citrus and the rubber tree. A pest problem that can arise with T-budding is an attack from red bud borer (*Resseliella oculiperda*), which lays its eggs under the bark of budded trees, especially apples. This problem can be controlled by sealing the tie with petroleum jelly or impregnating the ties with the essential oils of lavender (*Lavandula angustifolia*), which has been shown to reduce the infestation of buds by more than 95 percent.

Case study: T-budding of bush roses

T-budding gives very reliable results with bush roses and is the standard grafting technique used for this crop. The equipment required for T-budding is a very sharp T-budding knife, a sharpening stone, a second knife or secateurs for collecting bud-wood, rubber budding patches, clean damp hessian, and labels and a marking pen.

GROWING ROOTSTOCKS FROM SEED Several rootstocks are used for bush roses. *Rosa canina* was the traditional rootstock but it is very variable in terms of suckering, thorniness, and vigour. It has been improved as a rootstock by selecting varieties that are less variable. 'Inermis' is almost thornless and its bark "slips" late into the season, but the plant does have a tendency to sucker. 'Pfänder' is popular; it has a long straight neck, little suckering, grows well on light sandy soils, and can be budded late. It is prone to powdery mildew however. Other selections used to a limited extent

include 'Pollmers', a drought-resistant early budding variety prone to black spot, 'Schmid's Ideal', a good rootstock on lighter soils, and 'Heinsohn's Record', which is similar to 'Inermis'.

Rosa multiflora and selections are sometimes used but the rootstock that has been most popular for many years in the United Kingdom is *R.* 'Laxa'. This cultivar has few suckers, buds early, and is good on calcareous soils. It is raised from seed and was the subject of much research into pre-treatments so that it could be produced in the United Kingdom rather than being imported. This rootstock does not seem to perform as well elsewhere in Europe or the United States, where a hybrid Wichurana climbing rose 'Dr Huey' is the most popular rootstock.

If seed of *Rosa* 'Laxa' is sown immediately after collection in the autumn, then there may be some germination the first spring, about 3 percent, but the greatest germination will be after year two when up to 17 percent may germinate. There may be some further germination for a year, or even two, after this, but overall germination will be less than 25 percent from a species where seed viability at harvest is rarely below 90 percent. Germination is also spread over a long time and is not commercially practical to carry out.

The blade of a T-budding knife is slightly curved at the tip to make the T cuts easier and there is a tongue on the back of the blade to lift the bark after making the cuts.

141

Studies by J. B. Blundell at the University of North Wales, Bangor, and by researchers at the Horticulture Centre, Loughgall, Northern Ireland, were undertaken to try to improve the germination of *Rosa* 'Laxa' by pre-treatments prior to sowing. 'Laxa' has complex dormancy processes, but cleaned, dried, and properly stored seed ($<10°C$) should have a viability between 85 and 95 percent and be ready for the pre-treatments. It was found that a warm treatment at $21°C$ for 10–12 weeks, followed by cold stratification at $4°C$ for 12 weeks, would significantly improve the actual germination percentage. The final improvement in pre-treatments came with the use of sulphuric acid (acid scarification). Losses in seed viability occur during the warm treatment, as respiration rates will be high and energy stores will be used up. The time for the warm treatment can be reduced if some of the seed coat is removed by the use of acid.

Acid treatment is a dangerous process and should only be carried out with specialist equipment. The seed requires to be correctly dried after harvesting, since fresh rose seed is permeable to liquid and only fully develops and becomes impermeable with drying. Acid will also react violently with moisture, causing a rapid heat build up which must be avoided. The aim of the acid is to remove about two-thirds of the seed coat. Any more and there is the danger of the acid penetrating the seed coat and damaging the embryo. Since the thickness of the seed coat will vary between batches, test samples need to be treated first, removed at intervals, and cut open to assess the time taken to remove the required amount of the coat.

The main batch of seed is then acid treated for the required time. The temperature during the treatment should be between 40 and $50°C$ and not rise above $60°C$. The seed is stirred constantly to avoid hot spots occurring and to avoid a build up of charring (carbon caused by the acid reaction building up around the seed preventing further acid digestion). Following treatment, the seed is carefully moved to a large volume of water to neutralize the acid and stop further digestion of the seed coat. Following this, a warm treatment of only 4 or 5 weeks is required instead of 10–12 weeks. The warm treatment is followed by the 12 weeks of cold treatment. If carried out correctly, the acid, warm and cold treatments will give germination figures of 60–70 percent, and germination will occur in one year and therefore will be a viable crop to grow.

In the case of *Rosa* 'Laxa', it was recommended to sample the seed after 10 weeks. Seed can be cut to assess progress; the radicle prior to germination will appear bottle shaped with a greenish tinge. Some seed can be placed in moist sand in a petri dish at $20°C$, in light or dark. The optimum stage is where 3–5 percent of the seed will have chitted and the remainder is at, or near, the splitting stage; that is, the seed coat is opening as the embryo expands.

At this stage, the seed is separated from the medium and surface dried. It is then stored, spread out in wire mesh–based trays at $2–4°C$ to be sown within the week. It is also misted daily if required to prevent the seed drying out. If sowing is to be delayed, then the seed is air dried further until most of the seed is seen to be splitting. It is then stored at $10°C$ in linen bags until it is to be sown. Prior to sowing, it is soaked in water for 24 hours followed by chilling at $3–4°C$ for two days.

Rosa 'Laxa' should be sown in the spring to give a stand of 150–180 seeds per metre run of a 1.1 m-wide bed. The plants are lifted in the autumn and graded by neck diameter. The grades 3–5 mm and 5–8 mm are used for budding bush roses. The larger diameter should be used where soils or other growing conditions are not ideal, or in areas with relatively short growing seasons.

PREPARING ROOTSTOCKS To prepare the rootstock for grafting, the roots and top of the rootstocks are reduced in size to 100 mm and planted in the autumn. It is important to have a break between growing roses in any soil of at least four years to avoid problems of regrowth disease. Allow 1 m between rows and 300 mm between plants within rows. Do not plant too deeply but ridge up the soil around the necks of the rootstocks. The moist, dark conditions around the neck will improve the ability of the bark to slip. Remove the soil from around the neck just prior to budding. If the weather is dry, it is important to irrigate the rootstocks for two weeks prior to budding to ensure the bark will lift.

COLLECTING AND PREPARING THE SCION Collect strong, ripened shoots of the present year's growth once the buds are mature. An indication of this is that the first terminal flower has bloomed and is starting to go over. The spines on the stem should also snap off when pushed to the side, rather than bend. The stems, about 300 mm long with three or four usable buds, are cut from the parent plant and the leaves removed, leaving a short petiole. This can be done with secateurs or by holding the bud-wood towards you, placing the thumb of your knife hand behind a petiole, and cutting through the petiole to your thumb. The spines are then removed by bending them to the side to break them off. Cut off the soft tip of the bud-stick and bundle the sticks together to give, say, 100 buds per bundle. Remember to label each bundle carefully. The bundles should then be wrapped in moist hessian and placed in a polythene bag before being stored in a refrigerator or cool box until used.

Selected bud-wood is collected once the terminal flower has bloomed and the thorns will break, rather than bend, when pushed to the side.

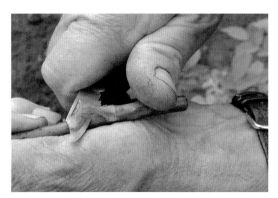

A healthy scion bud is removed from the bud-stick.

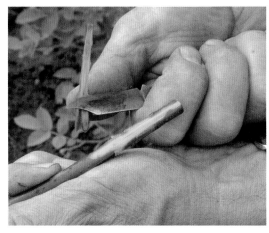

Final removal of bud. This is often cut off, rather than pulled as in the photograph.

The piece of wood behind the bud is often removed but this is not essential and is often left on the bud to save time.

To prepare the scion, the bud should be removed. You can start the cut from either above or below the selected bud. Whichever you do, the cut should start 25 mm from the bud and be trimmed to the same length once the bud is removed. Many propagators carefully remove the shield of wood from behind the bud. However, it has been shown that this does not significantly affect the bud take and so often the shield is now left, as this saves time.

T cut is made in the rootstock.

Insertion of bud behind the bark. If required, the top of the scion material can be trimmed to length so it fits into the top cut.

The bark is then pulled off the wood. It is important that this slips cleanly and does not tear.

The rubber patch is stretched over the bud and held in place with a metal staple.

MAKING THE CUTS Select a clean, smooth spot on the rootstock as low down as possible and make a vertical cut 40 mm long and just deep enough to penetrate the bark. Then make a horizontal cut 13 mm long at the top of the first cut to form the T. Lift back the bark on either side of the vertical to reveal the cambium layer beneath (it appears as green wood).

INSERTING THE BUD AND TYING-IN Insert the scion bud behind the bark, ensuring that you do not handle the cut surface as grease from your fingers may prevent the graft union from forming. The bud should be just below the cross cut. Trim away any surplus scion material above the T cut. Apply the bud patch over the bud to close the lifted bark and hold the bud securely.

The rootstock is headed back to just above the bud prior to new growth in the following spring. This will stimulate the scion bud to grow. Occasionally there may be shot buds—scion buds that have grown away in the previous summer after budding. In these cases, the scion shoot should also be cut back hard at the same time the rootstock is pruned.

Scion bud has taken and the rootstock can be headed back in the spring.

Chip budding

For many tree species T-budding can give significant losses from bud failure and, where buds do take, the subsequent growth can be variable. This situation led to the investigation of chip budding as an alternative method of budding trees.

In Europe, East Malling Research in Kent, England, led the research into chip budding and promoted it as having advantages over T-budding. Chip budding involves the removal of a section of rootstock and replacing it with a bud from the scion. Unlike in T-budding, the cut rootstock and scion are placed in direct contact, so there is close cambial contact leading to a faster graft union being formed. Trials of chip budding resulted in a reduction in losses from bud failure and often subsequent growth was more even, with a taller maiden growth being produced after the first growing season. The benefits were even seen in some species in the second year with more branches being produced compared with T-budding. These results have led to chip budding becoming the standard method of budding ornamental and fruit trees with most growers.

Chip budding has also been found to reduce the amount of union canker, because the spores cannot be inoculated behind the bark. In addition, since the bud has to be fully tied in to prevent desiccation, there is a lower incidence of red bud borer. Chip budding is also faster to carry out than T-budding and can be carried out over a longer period. While T-budding requires both the scion bud to be mature and the bark to slip and not tear when opened, chip budding can be carried out as soon as the bud is mature and later into the summer when the sap flow through the rootstock is reducing and the bark no longer slips for T-budding.

The correct timing is still important for chip budding. In the South of England, chip budding of *Hamamelis* has been tried. The optimum time for budding *Hamamelis* is in mid-August. Unfortunately, the scion buds are often not mature enough for budding until the end of the month. That is too late in the season to maintain rootstock growth, and therefore budding success is poor. Scion material could be grown under protection to produce the correct maturity of the bud earlier in August and improve bud take and subsequent growth.

Although chip budding has been around for a long time, it was not a commonly used technique mainly due to the difficulties in sealing the union quickly to prevent moisture loss. With the development of polythene tape for grafting, the tying and sealing of the chip bud became a

Japanese grafting techniques

Japan has contributed a number of grafting techniques to horticulture, although many of these were kept secret from the West for many years. In 1635, Japan entered a period of isolation known as Sakoku ("chained country"), which only ended in 1858. During this time, no foreigner could enter the country and no Japanese could leave or they faced the penalty of death.

For more than 200 years, the only western presence in Japan was at a trading station run by the Dutch on the man-made island of Dejima in Nagasaki Bay. In 1823, German physician Philipp Franz von Siebold (1796–1866) arrived at Dejima.

Siebold is a fascinating character who is not as well known as perhaps he should be. He was restricted to Dejima, but soon after arriving there, he cured a local official of an unspecified illness. This official had enough influence to enable Siebold to open a small practice outside Dejima and make house calls on Japanese patients. This led to contact with Japanese doctors and scientists who were keen to learn about advances in western science. Through these contacts and his ability to travel to his Japanese patients, Siebold was able to collect many specimens of native plants. Initially, he set up a small botanic garden on Dejima that contained

a thousand species and then he sent specimens back to the Netherlands. These included the first Hosta to arrive in Europe and the Japanese witch hazel (*Hamamelis japonica*). Many plants have been named in his honour.

One area that Siebold may well have visited was the nursery district of Angyo, now part of the cities of Kawaguchi and Saitama, north of Tokyo. This area has had nurseries since the 1600s and now contains the Omiya Bonsai Village built after the earthquake of 1923 when villages specializing in different crafts were established around Tokyo. Japanese propagators worked hard to develop new and better grafting techniques that differed from other propagators. Although this led to the development of many useful techniques, the propagators were also very secretive and a new technique would often only be passed on to a pupil of the propagator. Many techniques have therefore been lost in Japan with the death of the propagator and their pupil. However, Siebold was able to learn some of these grafting techniques during his travels and in 1828 he took these back to Europe. It is believed that chip budding originated on a nursery in the Angyo District and may have been one of the techniques seen by Siebold.

straightforward process. Although polythene tape will be suitable for all chip budding, growers have found that in some cases rubber ties can give better results. There does not seem to be an obvious reason why some varieties bud better using polythene and others with rubber. Matthews nursery uses rubber ties on *Sorbus* as it calluses quickly. At New Place Nurseries, however, they have found that it depends on individual species and cultivars and they cannot predict which one a plant will prefer. The advantage of rubber ties is that they are quicker to tie and will biodegrade within five or six weeks, so do not have to be removed by the propagator. The disadvantage of the rubber ties is that they are more expensive than polythene and not suitable for all species and cultivars. If in doubt, use polythene tape.

Case study: Chip budding of ornamental or fruit trees

The equipment required for chip budding is a very sharp knife (either a T-budding or grafting knife), a sharpening stone, a second knife or secateurs for collecting bud-wood, polythene or rubber budding tape 26 mm wide and 200 mm long, clean damp hessian, labels, and pencil or marker pen.

PREPARING ROOTSTOCKS Rootstocks of 6–8 mm should be planted in the spring using plants raised by mound layering or leafless winter cuttings. Prior to budding, the rootstocks are trimmed up beyond the budding height to give a clean stem. The height for budding can be between 50 and 350 mm, depending on the species. If the rootstock will be used to give benefits like vigour control, bud higher up the stem, about the length of the budding knife. Where the cuts on the rootstock are made depends on how the rootstocks have been planted. Normally, rows run north to south and the bud is put on the side of the prevailing wind—usually the west side. If

the rows run east to west, then the bud should be put on the north side of the rootstock. This will help the subsequent shoot to grow straighter, either due to the prevailing wind or due to growing towards the sun.

Select a straight section of the rootstock on the correct side and height. This will be an internodal cut; do not remove a bud from the rootstock. Make the first cut at the required height, cutting downward at an angle of 20–30 degrees to a depth of about 3 mm (Figure 6-1). Start the second cut about 40 mm above the first, keeping the knife at the same angle as the first cut. Use a toe-to-heel action, cutting inwards to nearly 3 mm and then straight downward in a controlled action to meet the first cut (Figure 6-2). Avoid going beyond and splitting the stem. The cut should be a slicing action and not a sawing (to-and-fro) action. Ensure that the second cut is deep enough to produce an inverted-U shape at the top and not an A shape. Remove and discard the cut piece of rootstock.

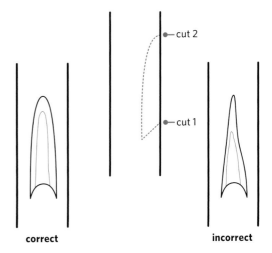

Figure 6-1. Removal of section from rootstock. Ensure that an inverted-U shape is produced and not an A shape at the top of the cut.

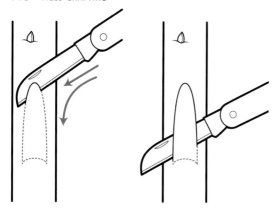

Figure 6-2. Following the bottom cut, 3 mm deep at a 20- to 30-degree angle, the top cut is made by starting at the toe of the blade. At the same angle as the first cut, go 3 mm into the wood and then slice straight downwards to join the first cut.

Bottom cut is made in the rootstock.

Second cut being made in the rootstock. Note the angle of the knife; it cuts in 3 mm at the top and is then pushed straight down to join the first cut.

Ensure the rootstock is healthy. This rootstock shows signs of staining once cut that may indicate the presence of *Phytophthora*.

COLLECTING AND PREPARING BUD-WOOD To prepare a bud-stick, collect shoots of a diameter similar to or slightly smaller than that of the rootstock. Ensure that the shoots are of the current year's growth and the wood is firm and mature with healthy buds. Avoid thin, soft flexible shoots, as these will not make successful grafts. Any diseased shoots should be left to be pruned and discarded later to avoid cross infection. Do not use your budding knife to collect bud-wood, but rather secateurs or another knife. It is best to collect material in the morning as transpiration will have been low during the night and so cells will be fully turgid. Leave at least one bud of this year's growth on the stock plant. There will be between five and ten buds on each shoot depending upon the species.

Once cut, the shoots should be put into a bucket of water or wrapped in damp hessian.

There are two methods of removing the leaves to prepare the bud-sticks from the collected material. The leaves can be cut off close to the bud, leaving as short a petiole as possible without damaging the bud. Do this with secateurs or by holding the bud-wood towards you, placing the thumb of your knife hand behind a petiole and cutting through the petiole to your thumb. Alternatively, some growers hold the tip of the bud-wood with one hand and slide the other hand down the shoot to remove all the leaves. In either case, the final bud-stick is prepared by cutting off the soft top of the stem and removing any damaged or malformed buds. Bundle the sticks together to give, say, 100 buds per bundle. Remember to label each bundle carefully. The bundles should then

Bud-wood at the correct stage of development for collection and preparation as bud-sticks.

be wrapped in moist hessian and placed in a polythene bag before being refrigerated or placed in a cool box until they are needed.

Chip budding is usually carried out in teams with one person preparing the cuts and one or two people tying-in the buds behind. This ensures that the buds are tied in promptly to avoid desiccation, although it is possible for one person to do the whole operation unaided. What is important is to have everything prepared before starting to bud. Ensure your knife is very sharp. The bud-sticks should be placed in an insulated bag that can be easily carried by the person budding. Plan to bud only one variety at a time and prepare labels to identify clearly the start and end points of each batch. Polythene or rubber tape 26 mm wide should be used and these should be cut to lengths of 200 mm prior to the start of budding.

REMOVING SCION BUDS Hold the bud-stick with its base towards your body. Select the bud to be removed and hold it so that it is about the width of the knife blade above your index finger. Make the first cut 20 mm below the bud, the width of the knife blade (Figure 6-3). Cut downward at an angle of 20 to 30 degrees to a depth of about 3 mm. Move the knife above the bud, again the width of the knife blade, so that the chip will be about 40 mm long. Keeping the knife at the same angle as in the first cut, cut inwards about 3 mm before cutting down towards the first cut (Figure 6-4). In this case, a heel-to-toe motion is used to give a slicing action. The cut should remain parallel down the stem; do not make the base of the cut deeper than the top and ensure that there is a U shape at the top of the chip and not an A shape. Be aware that there will be a point of resistance to the knife when cutting below the bud.

Once the bud is cut, lift it off the bud-stick using your thumb and knife blade. If the rootstock cut has been made first, then the bud is transferred to your non-knife hand and inserted into the rootstock. Commonly, the scion buds are prepared first and three or four are prepared and held in the budder's mouth while the rootstock cuts are prepared.

Make the first cut into the scion a knife-blade width below the bud.

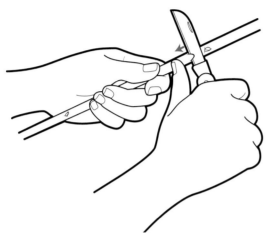

Figure 6-3. Make the first cut a knife-blade width below the bud.

Start the second cut by cutting 3 mm into the scion-wood and then coming straight down under the bud. Do not let the bud get thicker as you cut down. There is a natural point of resistance to the knife under the bud; take care that the knife does not slip.

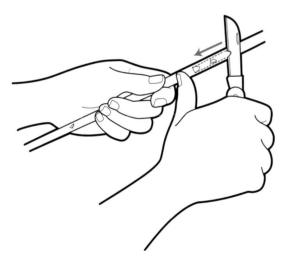

Figure 6-4. Start the second cut a knife-blade width above the bud. Cut in 3 mm at the 20- to 30-degree angle and then slice straight down to join the first cut.

INSERTING THE BUD-CHIP AND TYING-IN Do not handle the cut surface of the bud or allow it to drop on the ground, as any dirt or oil from fingers can prevent the union from forming. Set the bud so that it is held in place by the bottom angled cuts. If the scion overlaps the rootstock at the top, then lengthen the cut on the rootstock. If the scion is slightly thinner than the rootstock, then line it up so that the margin is equal on both sides. If it is a very small chip, then line it up on one side of the rootstock cut so that there is cambial contact on at least one side. The scion cut must not be broader than the rootstock (Figure 6-5).

To tie-in the graft, take a strip of the budding tape and start below the bud (Figure 6-6). Stretch the tape slightly to ensure that pressure is applied to the bud-chip, and place the stretched end round the bottom of the cut. Ensure that the cut is well covered and pass the longer end of the tape round the back of the rootstock. Keep the tape tight and go slightly lower around the back of the rootstock and then up to catch the short end of the tape. Wrap the tape up the stem, ensuring the edges overlap and the pressure is maintained (Figure 6-7). Only one thickness of tape should cover the actual bud, and often the pressure is released slightly to ensure the bud is not damaged. Once all the cut surfaces are covered, finish the tie with a half hitch: form a loop round your finger and put the end of the tape under the loop. It is important to continue tying-in the same direction while

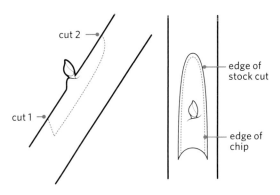

Figure 6-5. The bud should fit neatly with the rootstock cut, with edge of the rootstock just visible behind the scion bud.

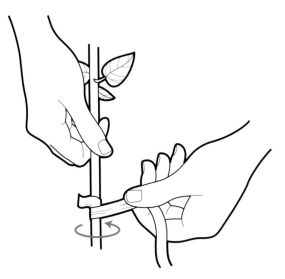

Figure 6-6. Start the tie at the bottom of the chip. Overlap one end of the tape to hold it in place. You should just start to stretch the tape to get the correct tension.

The chip bud placed on the rootstock.

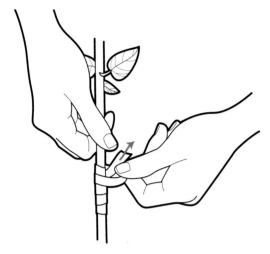

Figure 6-7. Ensure all the cut surfaces are covered and the polythene tape is overlapped up the chip. Tie off at the top by forming a loop in the tape and putting the end through and pulling tight. Do this in the same direction as you were tying so that the tension is not lost.

Bud being tied on the rootstock.

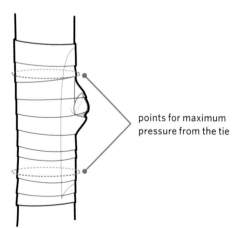

points for maximum pressure from the tie

Figure 6-8. It is most important that pressure be applied at the top and bottom of the bud. If the bud is large, do not cover it but do make sure all the cut surfaces are covered.

Completed tie with polythene tape.

Completed tie with rubber tape. In this case, the bud has also been left exposed.

doing this, and put the tape through the loop so the end is pointing upwards. Ensure the tie is tight and secure before moving on (Figure 6-8).

Prominent buds may be damaged if the tie covers them. In this case, the tape is passed above and below the bud as closely as possible so that the cut surfaces at the side are covered, but the bud itself is left exposed.

After four to six weeks, the ties need to be removed, unless rubber ties have been used. With a knife, cut the half hitch knot and unwind the tape. Do not cut the bark when doing this. Cut away from the rootstock when removing the tie. The following spring, cut the rootstock back just above the scion bud. Tie the subsequent scion growth to a cane to produce a straight stem to the new tree.

The rootstock is left attached over the winter.

In early spring, before bud break, the rootstock is headed back to just above the bud.

Winter grafting

Rootstocks grown in the field for a season can be grafted in late winter. This means that any rootstocks where a bud failed to take the previous summer will not have been wasted. Winter grafting is possible because tree species that are budded in the summer, like apples and rowans (*Sorbus*), will form a graft union at low temperatures. A whip-and-tongue graft is commonly used for this type of graft as it holds the union together securely at a time when rain and strong winds will be common. Winter grafting is also an option on a small scale where growing plants in containers for bench grafting is not feasible.

7 Vegetable Grafting

GRAFTING VEGETABLE PLANTS HAS BEEN known for a long time. In an article of 1875 for the Royal Horticulture Society magazine *The Garden* Frederick Burbidge wrote the following:

> In the *Gardeners' and Land Stewards Journal* 1847 (p. 85) is given a short account of an experiment in which a Tomato scion was grafted upon the stem of a Potato, and the scion developed its fruit and the stock formed tubers.

Burbidge further mentioned herbaceous, cucumber, and potato grafting in his book *Cultivated Plants—Their Propagation and Improvement*. In the United States, grafting of vegetable crops was also being investigated seriously in the late nineteenth century. In 1890, James Edward Rice was studying for a Bachelor of Science in Agriculture at Cornell University. For his thesis, he studied herbaceous grafting. According to reports, the work was "planned and wholly carried out in the most careful manner by Mr. Rice" and it was thought worthy of publication. Rice carried out a number of different grafting techniques on geraniums, chrysanthemums, and coleus, as well as tomato, pepper, and aubergine. He tried a number of reciprocal graft combinations, including pepper with tomato and pepper with aubergine. He also showed that tomato grafted onto potato produced and

> bore good tomatoes above and good potatoes beneath, even though no sprouts of the potato were allowed to grow, but that mealy bugs were particularly troublesome upon these grafted plants, for they delighted to crawl under the bandages and suck juices from the wounded surfaces.

Despite these efforts, grafting of vegetables appears to have remained a curiosity in Europe and the United States until well into the twentieth century following its adoption on a large scale by the Japanese. Japan was facing the problem of an increasing population with limited space in a mountainous country for housing and food production. Where land use is not an issue, then crop rotation can be practised and the risk of soil-borne diseases building up is reduced. In Japan, however, land use is very intensive and crop rotations are more difficult to carry out. In the early 1920s, soil-borne diseases caused by *Fusarium*, *Verticillium*, and *Pseudomonas* organisms were having a significant effect on vegetable yields.

It seems to have been a Japanese watermelon (*Citrullus lanatus*) farmer who first tried grafting to overcome this problem, using a rootstock of squash (*Cucurbita moschata*), which is resistant to *Fusarium*. The work of this farmer led to grafting trials at regional agricultural experimental stations across Japan, and the success of that work led to the rapid dissemination of grafting techniques throughout Japan and into Korea. By the 1930s, it was common practice to graft watermelon, cucumber (*Cucumis sativus*), Oriental melons (*Cucumis melo*), tomato (*Solanum lycopersicum*), and aubergine (*Solanum melongena*). These crops were also investigated for other potential benefits of grafting beyond reducing the problems of soil-borne diseases. Among these additional benefits were enhancing water and nutrient uptake, increasing plant vigour, extending the harvest season, and increasing the tolerance of plants to low temperatures and saline and wet soils. By the end of the twentieth century, about 651 million grafted vegetable plants were being grown each year in Japan, including more than 95 percent of watermelons, most greenhouse cucumbers, and up to 30 percent of outdoor cucumbers, and most Oriental melons.

Benefits of grafting vegetables

Even where soil-borne diseases are not a problem, it can be advantageous to use a particular rootstock for other reasons, such as to increase yields and length of cropping. Breeding rootstocks for resistance or tolerance to a range of environmental factors is separate from breeding for fruiting characteristics of the scion.

Interspecific rootstocks created by the plant breeder are particularly valuable to introduce characteristics to the rootstock while maintaining its compatibility with the scion. The cut stems of grafted cucurbitaceous crops, such as watermelon, pumpkin, and Oriental melon, often show increased amounts of xylem sap compared to the cut stems of self-seeded plants. This xylem sap contains high concentrations of minerals, organic substances, and plant hormones such as cytokinins and gibberellins. The ability to absorb water nutrients and hormones more efficiently may be due to the production of a greater volume of roots that is often seen in rootstock cultivars. In aubergine, yield has been shown to correlate positively with the amount of xylem sap exudate. The greater water and nutrient uptake is also believed to increase the cropping period and is particularly important in protected crops.

Although the size of fruits and overall yield may be greater with grafted plants, there is some debate as to the effect of grafting on quality. Some of the important quality measures in cucurbits are Brix degrees, firmness, rind thickness, rind colour, and fruit shape. Abnormal fruit quality has been reported in some grafted plants. Oriental melon can have reduced fruit soluble solids content, persistent green stripe in the skin, fruit fermentation, fibrous flesh, and off-taste. Watermelon can have reduced Brix levels, increased number of yellowish bands in the flesh, insipid taste, poor texture (more fibres), and decreased firmness. However, other reports show a positive effect of grafting on watermelon quality, including an increase in fruit firmness and Brix readings. The effects on quality may be due, in part, to the rootstock used. It has been shown that watermelon fruit from scions grafted onto *Citrullus maxima × C. moschata* rootstock had no difference in soluble solids concentration compared with the control that had not been grafted.

There may also be an effect on maturity of fruit at harvest. Unlike watermelon, cucumber is picked immature and there have been fewer negative reports of grafted plants affecting quality. In a few studies, however, it has been shown that cucumbers do show some variation in factors like shape, skin and flesh colour, texture, skin smoothness, and rind thickness when different rootstocks have been used. Therefore, rootstock selection needs to be done with care.

To complicate matters, not only will cultivar-specific characteristics of the scion and those of the rootstock affect quality, but the subsequent cultivation of the plants will also affect quality. Changing cultivation recommendations for the growing crop may possibly minimize any detrimental effects on quality; this is principally by reducing fertilizer input and avoiding excessive irrigation to prevent excessive vegetative growth and improve the appearance of the fruit. Specific nutrient treatments, for example, foliar application of calcium and reduced nitrogen levels, have also been shown to reduce internal decay in some fruit.

Low-temperature tolerance in rootstocks was also found to be a benefit of grafting. The fig leaf gourd (*Cucurbita ficifolia*) used as a rootstock for cucumber can be used for early spring growth in unheated greenhouses. In areas of the Middle East, grafting is commonly used with salt-tolerant rootstocks. Rootstocks that are tolerant to flooding may also be valuable in some areas. Interspecific grafting of solenaceous vegetables can give interesting combinations too. For example, a tomato scion can be grafted onto potato (*Solanum tuberosum*) to give a double crop. This novel crop is something that the amateur gardener may find of interest to try, but would not be a commercial approach to growing these crops.

In summary, the major benefits of using grafted seedlings are to achieve resistance to soil-borne diseases and nematodes, to increase yield and quality, and to improve the physiology of plants making them more adaptable to harsh environments. In Asia, the main driver to the development of grafted crops has been the need to overcome soil-borne diseases, while in Europe, increase in yield and quality has led to a return of the use of grafted plants. In other countries, like in the Middle East, the adaptability to harsh environments has been important in making grafted crops popular.

Table 5. Rootstock, major grafting methods, and purpose of grafting for selected vegetables in Japan and South Korea.

FOOD CROP	ROOTSTOCK SPECIES*	Hole insertion	Approach grafting	Cleft grafting	Fusarium wilt control	Growth promotion	Low temperature tolerance	Growth period extension	Nematode resistance	Bacterial wilt control	Virus infection reduction
Citrullus lanatus (watermelon)	*Lagenaria siceraria* (gourd)	•			•	•					
	Interspecific hybrids**	•	•		•	•	•				
	Benincasa hispida (wax gourd)	•		•	•	•					
	Cucurbita pepo (pumpkin)		•	•	•	•	•				
	Cucurbita moschata (squash)	•	•		•	•	•				
	Sicyos angulatus (star-cucumber)		•						•		
Cucumis sativus (cucumber)	*Cucurbita ficifolia* (fig leaf gourd)		•		•	•	•				
	Interspecific hybrids**	•	•		•	•	•				
	F1 (*Cucurbita* × *C. moschata*)		•		•	•			•		
	Cucumis sativus (cucumber)		•		•	•					
	Sicyos angulatus (star-cucumber)		•			•				•	
Cucumis melo (Oriental melon)***	Interspecific hybrids**		•		•	•	•				
	Cucurbita moschata (squash)		•		•	•	•				
	Cucumis melo (Oriental melon)			•		•	•				

Source: Adapted from J.-M. Lee (1994).
*Named cultivars are available from seed suppliers.
**Many interspecific hybrids are commonly obtained by fertilized ovule culture in vitro.
****Cucumis melo* (Oriental melon) is grown extensively in Japan and Korea.

FOOD CROP	ROOTSTOCK SPECIES*	GRAFTING METHOD			PURPOSE							
		Hole insertion	Approach grafting	Cleft grafting	Fusarium wilt control	Growth promotion	Low temperature tolerance	Growth period extension	Nematode resistance	Bacterial wilt control	Virus infection reduction
Cucumis (other melons)	*Cucumis melo* (Oriental melon)		●	●	●						
Solanum lycopersicum (tomato)	*Solanum pimpinellifolium* (currant tomato)			●					●		
	Solanum habrochaites (wild tomato)			●					●		
	Solanum lycopersicum (tomato)			●					●		
Solanum melongena (aubergine)	*Solanum integrifolium* (scarlet eggplant)		●	●						●	
	Solanum torvum (turkey berry)		●	●							●

Techniques for grafting

Three main methods are used in the grafting of vegetables: hole insertion, approach grafting, and cleft grafting.

Hole insertion

Hole insertion is used for watermelons because of their small seedling size compared to the size of the gourd or squash rootstock, but it can also be used for melons and cucumber on appropriate rootstocks. It can be done with or without roots on the stock. Unrooted stocks are easier to handle and produce strong adventitious roots. This method makes a strong graft, but great care is required to control the environment during healing.

The scion seedlings are ready for grafting when the first true leaf is visible by about 5 mm,

normally seven to ten days after sowing. The rootstock hypocotyl (the part of the stem of an embryo plant beneath the stalks of the seed leaves, or cotyledons, and directly above the root) should be well elongated with the first true leaf about 19 mm long, seven to fourteen days after sowing.

About two hours prior to grafting, cut the rootstock at the base to give a 5- to 6-cm stem containing the cotyledon and first true leaves. This allows the rootstock to lose a little turgor, which in turn reduces the risk of cracking during grafting. The true leaf is then carefully removed from the rootstock just above the cotyledons. To avoid shoots developing from the rootstock, the tiny axillary buds at the base of the petiole need to be removed. A pointed scalpel is used to cut at an angle from each side of the cotyledons. This also produces the "hole" for the scion.

The scion is cut to a length of 1.5 cm, and two wedge cuts, 7–10 mm long, are made at 90 degrees to the direction of the cotyledon leaves. The pointed cuts of the scion are inserted into the hole of the rootstock so that the cotyledon leaves are at a 90-degree angle to each other. No clips or ties are required, and the completed graft is inserted into a well-watered rooting medium. Do not let the rootstock wilt too much prior to grafting, and mist the grafts with a fine spray of water periodically before they go into the rooting environment.

The grafts need to be placed in 100 percent humidity at 28–29°C and kept in complete darkness for the first 24 hours. This could be in a fog unit or under a polythene tent. Light is introduced after 24 hours but heavy shading should be used so that it is just above the light compensation point. After five or six days, the humidity is lowered to begin the weaning process. After seven days, the plants are removed from the propagation area to the greenhouse where they will require further shading to continue the weaning process back to growing in full light conditions.

Approach grafting

Approach grafting is carried out when opposing and complementary notches are cut in the stem of the rootstock and scion. The complementary notches are fitted together and held with a spring clip or tape. Since the scion stays attached to its roots during the graft process, a high success rate should be achieved.

Approach grafting can also be used when scion and stock are slightly different sizes, but is a slow method if doing large numbers of grafts. The scion and stock should be at the two- or three-true-leaf stage before cuts are made halfway across the hypocotyl at a 60-degree angle. This will be a downward cut for the rootstock and an upward cut for the scion. Interlink the two "tongues"

produced by the cuts and hold them together with a grafting clip.

The two plants are then potted into a container and, after seven days, once the graft union has healed, remove the graft clip and cut the root system from the scion plant. Then the shoot is removed from the rootstock plant.

Cleft grafting

Cleft grafting is carried out when the plants are at the slightly larger two- to four-true-leaf stage, generally five to seven days after sowing. It involves cutting the rootstock halfway between the cotyledons and first true leaves. Remove the scion at a similar height to the rootstock. Trimming the leaves is often recommended to reduce the water stress on the scion while a graft union is forming. A V-shaped cut is made in the stem of the scion. The scion is then inserted into the rootstock, which has a vertical slice cut down the centre of the stem. The rootstock and scion are then held together by a spring graft clip while the graft union forms. The grafts are placed in a 90 percent humid environment under a polythene tent, with 50 percent shade at 18–21°C daytime temperature and 16–18°C night temperature for seven days. Capillary watering should be carried out to ensure the graft union does not become wet. After one week, reduce the humidity to 50 percent, remove the shade, but keep the temperature and watering regimes the same. After two weeks, the grafts should be weaned back to greenhouse conditions for growing on.

Commercial production of vegetables

It was probably after the Second World War that European growers learned about the Japanese use of grafted vegetables. In 1947, the grafting of vegetable crops for commercial production began in Europe. Because salad crops like tomato and cucumber must be grown under protection in Northern Europe, greenhouse growers encountered the same problems of soil-borne disease due to intensive cropping that the Japanese had encountered.

Breeding to introduce disease resistance has been used for salad crops, but this is not feasible in all cases. For example, the genetic trait in tomatoes for resistance to bacterial wilt (*Ralstonia solanacearum*) is closely linked to a negative trait for small fruit size. The main technique to prevent soil-borne diseases building up in a glasshouse was, therefore, to sterilize the soil between crops. This was done by steam, which required high-pressure boilers and was expensive in fuel use. Alternatively, methyl bromide, a colourless, tasteless, odourless, and nonflammable gas was used as a very effective soil sterilant. Unfortunately, this gas is also damaging to human health and the ozone, and although commonly used in the 50s and 60s, has been withdrawn from use.

In 1947, Dutch cucumber growers first began grafting cucumbers as an alternative to soil sterilization. Rootstock resistant to soil-borne diseases could be bred without the problem of also having to have good fruit qualities. It took until 1962, however, before grafted tomatoes began to be used commercially. At this time, tomato rootstock seed was even sold to amateur gardeners, allowing them to do their own grafting, but the practice does not seem to have caught on and never became popular with the amateur.

Around the time grafted tomatoes began to be used commercially, however, novel soilless methods of growing tomatoes were being developed. At the Glasshouse Crops Research Institute in Southern England, Allan Cooper developed the nutrient film technique (NFT). This technique takes plants out of the border soil and grows the roots in a thin film of water and nutrients that are circulated through plastic channels, keeping the roots oxygenated. This system was followed in the early 1970s by the introduction of the grow bag from Fisons Horticulture. The grow bag also took the plants from the soil but, instead of just water, the plants went into a peat-based growing medium that was irrigated and fed through a drip system. Further systems were developed using the likes of rockwool, perlite, and coir, which support the roots and provide a medium for the supply of water, nutrient, and oxygen. These systems enabled growers to have control that is more precise over the growth of the plants and reduced the risk of soil-borne diseases by taking the plants out of the soil. Such developments meant that there was not a need for grafted plants, and the practice of grafting vegetable plants in Europe almost ceased. This was partly due to the reduced risk from soil-borne diseases in the new methods of production, and also because of the cost of grafting plants.

It was not until the 1990s that grafting salad vegetables began to be used again. This revived interest was partly due to three factors: (1) seed companies marketing new rootstock cultivars that had desirable traits to increase yields significantly and, (2) researchers exchanging information on the benefits of rootstocks, and (3) the introduction of commercial production techniques for the propagation of grafted seedlings.

The main glasshouse crops grafted in Europe at present are tomatoes and aubergine, although bell pepper and cucumber can also be grafted. Rootstocks for European greenhouses have often been developed separately from those commonly used in Japan. The principle tomato rootstocks are F1 interspecific hybrids of *Solanum lycopersicum* crossed with *S. habrochaites*. These hybrids are resistant to many of the potential soil-borne diseases that can affect tomatoes and also produce strong, vigorous root systems that will increase and extend the harvest season. The increased vigour enables two main stems to be trained from each rootstock, which in turn can reduce the number of plants in a house by half, and thus reduce the extra cost of using grafted plants compared to seed-raised plants. The same hybrid tomato rootstocks are used for glasshouse aubergines in Northern Europe. This is probably due to the amount of breeding carried out to improve tomato rootstocks and the familiarity of growers with tomato rootstocks. In Italy, however, where a more diverse range of aubergines is grown, the tomato rootstock has been shown to have a negative effect on yield and quality on some cultivars. Other *Solanum* species, such as *S. torvum* and *S. sisymbriifolium*, reportedly give promising results as rootstocks on a wide range of aubergine cultivars.

Glasshouse producers of vegetables tend to keep to a known stock even though plant breeders are selecting and releasing new cultivars. This is because of the need to adjust cultural practices of irrigation and nutrition to suit a particular rootstock. Once growers have become familiar with managing one cultivar, there will have to be significant benefits in an alternative to get them to change and learn how to grow on a new rootstock.

There are good opportunities for amateur gardeners to use grafted vegetable plants. Grafted plants are available from a number of suppliers of young plants. Seed of rootstock cultivars is also available from seed companies, making it possible to graft one's own plants. If using grow bags in a greenhouse, then grafted plants can give higher yields over a longer period. If cold-tolerant rootstocks are available, then it may be possible to start the crop earlier in the year or reduce heating costs if using a heated glasshouse. Outdoor

tomatoes may also benefit from these advantages, and there will be the benefits of disease resistance of the rootstocks. If growing organically, then grafted plants will probably be even more beneficial. If growing in the border soil of a greenhouse, the problems of soil-borne disease build up will be reduced by using the correct rootstocks, since crop rotations are not really an option in these circumstances.

Whip graft of tomato, sweet pepper, and aubergine

In Europe, the most common graft used by commercial propagators of tomatoes is the whip graft. Sweet peppers and aubergine are grafted the same way (as described below), but the number of days from sowing to grafting is longer. It will take grafts of peppers and aubergine a few more days to establish and the plants to wean off. Aubergine is grafted onto tomato rootstocks, while sweet pepper is grafted onto sweet pepper rootstock.

When grafting tomatoes, both the rootstock and scion are cut below the cotyledon leaves at an angle of 45 degrees. The rootstock and scion are then joined with a silicone clip. The clip is nipped to open it up and slid halfway over the rootstock. The scion stem is then slid in from the top. The clip is released to allow it to hold the stems of the scion and rootstock in place. Care must be taken to ensure the exposed stem angles match perfectly and no debris is lodged in the union.

It is important that the diameter of the rootstock and scion stems closely match. The speed of germination and early seedling growth, therefore, must be considered when choosing the relative sowing dates of the rootstock and scion, to ensure that the stem diameters match when grafted. This means that the sowing dates of scion and rootstock can be different. Generally, most are ready to be grafted approximately 11 days after sowing. If it looks like the scion and rootstock will not match, then the faster-growing plant can be held back by growing it at a cooler temperature until it is the correct size. It is important to ensure that the growing media of both scion and rootstock are adequately watered prior to grafting, and the rootstock continues to be adequately watered during the period of graft union and then weaning.

The blade used for making the cuts must be very sharp to ensure a precise and clean cut. For vegetable plants, a razor blade is usually used. The blade must be changed as soon as it becomes even slightly blunted so that the cuts do not become ragged. Hygiene is also very important throughout the grafting operation. Work surfaces should be disinfected, gloves worn, and any tools cleaned between different varieties to minimize the risk of passing an infection from one variety to the other.

Immediately after the scion and rootstock have been joined, the plants should be misted with a

A grafted tomato plant.

Attaching the silicone clip onto a tomato plant to complete a graft.

Commercial glasshouse producing tomatoes from grafted plants. Note two stems taken from the graft.

fine water spray, placed in an area of very high humidity, and shielded from harsh sunlight. This can be under either a mist propagation unit with frequent misting, or polythene. If the latter, then care must be taken to keep the polythene off the plants where it could disturb the delicate graft. Humidity should be maintained at over 90%. During this period, it is common for the leaves of the scion to wilt severely.

The time that it takes for a union to form will vary from variety to variety but is usually two or three days. A sign that the union has formed is when the scion foliage is no longer wilting. To test a plant, gently pull on the scion to see if it has fused to the rootstock. Once the scion and rootstock have started to fuse together, the process of weaning the plants off the very high humidity environment should begin.

The aim of weaning is to gradually reduce the humidity and increase the light. One way to accomplish this is by reducing the frequency of misting under a mist unit. Another way is to open the polythene tent gradually and reduce the shade levels. Plants kept in high humidity for too long can end up with several problems. First, more sap than can be transpired by the leaves in the high humidity is pumped through the roots, leading to guttation (the exudation of drops sap on the tips or edges of leaves). Second, plants will stretch (etiolate) under the low light and high humidity. Third, risk of disease increases in the high humidity.

Grafted aubergine on a tomato rootstock.

Close up of graft union of aubergine scion and tomato rootstock.

It is important to start to make the plants work on their own as soon as possible to encourage the union to establish as quickly as possible. If the plants are too comfortable, then the union will take place more slowly.

Modified whip graft of cucumber

The method of grafting for cucumbers differs a little from that of grafting tomatoes. The stem of the rootstock is severed from the roots leaving 70 mm of stem below the cotyledon leaves. A 10 mm long cut at a 45-degree angle is made to the rootstock near the growing point. The aim is to cut off the rootstock's growing point and one of the cotyledons, leaving the other attached. The scion is cut to a length of 30 mm below the cotyledons and a 10 mm long cut at a 45-degree angle is made at the base. The rootstock and scion are matched up and the graft union secured with a grafting clip. This results in a plant that has three cotyledons. The one from the rootstock provides photosynthates while the rootstock roots and graft union forms. The stem of the rootstock is inserted 30 mm in to a rooting medium. The union should form in 7 days if kept at a high humidity under polythene with 70% shade. A good grafter can prepare 80–100 grafts per hour. The rootstock for cucumber is a hybrid of the family Cucurbitaceae.

8 Grafting Cactus

THE GRAFTING OF CACTI IS RELATIVELY straightforward and I am very grateful to Brian McDonough of the British Cactus and Succulent Society, Glasgow Branch, for much of the information in this section. There are multiple reasons to graft cacti, and this chapter will look at six of them. It will also examine the advantages and disadvantages of several rootstock species and describe an example of a cactus graft.

Reasons to graft cacti

Some types of cacti lack chlorophyll and cannot photosynthesize. These mutations can occur in seedlings of *Gymnocalycium mihanovichii*, known as moon cactus. Unless the seedlings are grafted onto another cactus that contains chlorophyll, they will not survive. This is the first reason for grafting cacti.

A second is to propagate cacti with cristate or monstrose forms of growth that can be difficult to root from cuttings. Such forms occur where there is a mutation of the apical meristem. In cristate mutations, a fanlike or crested growth occurs, while in monstrose growth, apical dominance is lost and a jumbled growth of knobby or lumpy appearance occurs. Grafting these unusual forms is the only way to produce additional plants.

Reason number three is that some pincushion cacti (*Mammillaria*) are prone to rotting on their own roots. To rescue these plants, indeed to cultivate them at all, often requires them to be grafted.

A fourth reason is to accelerate the growth of very slow growing species, such as *Aztekium ritteri* and *A. hintonii*. These can take decades to flower from seed, but will flower within a few years if grafted onto a vigorous rootstock. *Aztekium ritteri* was only discovered in 1929 in Mexico and was thought to be a genus with only one species until 1991 when *A. hintonii* was found. A further species, *A. valdesii*, was found in 2011 although its name has yet to be officially recognized. Grafting these rare slow growers allows more people to cultivate them.

Grafting may also be a way of increasing the numbers of choice material quicker than other methods of propagation. Some genera of cacti remain solitary while others naturally offset. The offsetting plants will be more disposed to offsetting when grafted and, in turn, these new offsets can be grafted. The solitary plants, with an apical growing point, will just continue to grow in this manner. However, if the apical growing point is removed, then the plant will begin to produce plantlets from the areoles. These can, in turn, be removed and rooted, or re-grafted.

The areole is a unique feature of cacti and gives rise to the spines. The fact that the spines arise from the areoles and not directly from the stem enables cacti to cover themselves with the defensive spines more efficiently than other plant species.

Because grafting of cacti accelerates their growth and reduces the time to flowering, plant breeders use it to speed up the process of hybridization and selection. This is the sixth situation in which grafting cacti is very desirable.

Rootstock species

There appears to be very few problems with incompatibility between different genera of cacti. The choice of rootstocks is, therefore, quite large, but it is necessary to consider the advantages and disadvantages of each plant to get a suitable rootstock for your purpose. In general, very fast growing rootstocks are usually short-lived. They tend to be from tropical climates and are best used for short-term gain, for example, getting seedlings to flower quickly. Slower growing rootstocks usually live longer and keep the scion more typical in form. In addition, larger or taller rootstocks will last longer than shorter stock. Some of the cacti used as rootstocks are described here. The list is not comprehensive.

Echinopsis species
Echinopsis species are globular to short cylindrical cacti from the mountains of South America. They are easily propagated from offsets. Because they are not overly vigorous, stocks usually remain true to type. The species are generally cold hardy. Grafting low on the rootstock may lead to it being hidden by the scion. The main disadvantage of using these species is that beheading usually causes the rootstock to produce many offsets, taking the strength away from the scion.

Harrisia jusbertii
Although the genus *Harrisia* is South American, the habitat of *H. jusbertii* is not known. It is believed to be a naturally occurring hybrid between *Harrisia* and *Echinopsis*. Whatever its taxonomic standing,

H. jusbertii is good for large seedlings and small offsets. It also promotes good scion growth. Disadvantages of using it are that it is a relatively short-lived species and needs warm winter temperatures, ideally above 8°C. It is less readily available from nurseries than other rootstock species and is perhaps best grown from seed specifically for grafting purposes.

Hylocereus species

Hylocereus species are tropical, clambering plants. They are easy to propagate and fast growing given enough warmth. They are good for grafting small offsets and medium-sized seedlings, and promote quick scion growth. On the negative side, they tend to be short-lived in temperate climates, requiring a minimum of 15°C to keep the rootstock alive. Since the genus is tropical, the species do not experience dormant periods in nature and must be kept growing year-round. This situation is not ideal for scions that would normally experience a dormant period. Scions also tend to produce atypical or bloated growth if they do survive. It is probable that most grafted cacti sold in Europe originate from the Far East and have been grafted on to *Hylocereus* rootstocks. Often these plants will rot and die in the first winter in the United Kingdom due to their requirement for relatively high temperatures.

Myrtillocactus geometrizans

Myrtillocactus geometrizans is a moderately fast-growing, columnar desert plant from Mexico. Young seed-grown plants are usually used as rootstocks for ease of production. This species is good for small- to medium-sized offsets; the scions are fast growing, and the rootstocks will withstand a period of dormancy. A disadvantage of using this tropical species as a rootstock is that it needs to be kept at a minimum of 10°C over winter. In the cool, damp climates such as that of the United

Kingdom, this cactus does not do well even in the greenhouse in winter. Plants will be marked with cold spots if kept too cold, although they may struggle to grow for a few more years.

Opuntia compressa

Opuntia compressa, a frost hardy prickly pear type of cactus from North America, makes an ideal rootstock for cold tolerant plants growing in northern climates. It is good for grafting small offsets and seedlings. In addition, because it has jointed pads, its vascular bundles are arranged in an elongated sausage shape rather than a ring. This arrangement facilitates the grafting of multiple seedlings onto one rootstock if required.

Pereskiopsis species

Pereskiopsis is a genus of fast-growing, tropical, shrubby cacti that grow into large bushes. These species are extremely useful for grafting small (sometimes just a few days old) to medium-sized seedlings. Rootstocks are easily propagated from cuttings and are fast growing. Scions develop quickly on these rootstocks. When grafting very young seedlings, it is best to cut with a razor blade so as not to bruise the tissue at the point of the cut. The scion can be slid off the razor blade onto the prepared stock.

Disadvantages of using *Pereskiopsis* species are that they are short-lived and not suitable for medium to larger scions. They need warm winter temperatures, ideally above 10°C, and tropical conditions for propagation.

Selenicereus species

Known as queen of the night or nightblooming cereus (although, like many common names, these common names are also used for other species), *Selenicereus* species are fast-growing, clambering and dry forest tropical cacti native to Central America, the Antilles, and Mexico. They are vigorous growers but not often used as a rootstock. Advantages of using them are that they are good for medium to large seedlings and small offsets, and they promote fast growth of the scion. The primary disadvantage is that these are relatively short-lived plants requiring warm winter temperatures above 10°C.

Trichocereus species

Trichocereus species are moderately fast growing cacti from mountainous South America. Either young plants or large plants can be used as rootstocks. The commonly used species is *T. spachianus*, but *T. macrogonus*, *T. bridgesii*, and *T. pachanoi* also perform well. Advantages of using these species are that rootstocks can easily be grown from seed or cuttings and are good for small- to medium-sized offsets on young rootstocks. Bigger offsets and larger scions are grafted onto larger rootstocks. *Trichocereus* rootstocks can be used with many different species of scion. They are also capable of withstanding low temperatures and can survive periods of dormancy, and so are very useful for growers in cooler, temperate climates. Generally, *Trichocereus* rootstocks live longer than many other rootstocks. The main disadvantage of using *T. spachianus* as a rootstock is that its vigorous habit can cause the scion to which it is grafted to become slightly bloated. *Trichocereus macrogonus* and *T. bridgesii* rootstocks have given less bloated results.

Do's and don'ts

Grafting is carried out during the growing season in late spring and summer when there is active growth. Clean cuts with a scalpel or craft knife blade are used to make the initial cut through the tough epidermis and spines. A razor blade is then used to clean up both surfaces and give a good, flat surface. It is important to remove any damaged or diseased material and take care not to cause pressure damage when making the cuts.

The size of the scion and rootstock should match—about 30 mm in diameter is recommended. To prepare the rootstock, cleanly cut off the top slightly above the required final height so the cut surface can be cleaned up and bevelled if required. To prepare the scion, remove a small section of stem. If the diameters of the rootstock and scion do not match, then bevel cut the shoulders of the larger piece to prevent the puckering that occurs as the cut surfaces dry out, and which can cause the union to separate.

In cacti, the vascular cambium is easily seen as a distinct ring on the surface of the cuts and at least some of this has to be matched up when the scion is placed on the rootstock. It may be noticed that even where the body diameters are the same, the vascular material may be of different diameters. This is not a problem as long as there is some overlap between the scion and rootstock.

To hold the graft together, use rubber bands that stretch round the bottom of the pot containing the rootstock and over the scion, so that pressure is applied on the scion. To ensure the rubber bands stay in place, position them at right angles to each other. Unlike grafts of woody plants, cacti grafts do not require to be sealed or kept in a humid environment; they are just grown as normal. In the summer, the elastic bands can be removed carefully after two or three weeks. If they are left on much longer, there is a danger that the growing tip of the scion can be damaged.

In the case of grafting very young rootstocks onto *Pereskiopsis* species, rubber bands cannot be used as the seedlings are very fragile. With these very small scions, the surface tension and sap should be enough to keep the graft together while the union is forming, although some growers use superglue around the outside of the graft to keep it in place.

Grafted cacti can grow for years. However, after a time, the rootstock tissues will cork over from old age and the scion will need to be grafted again onto a new rootstock, as the corky growth will prevent nutrients getting to the scion.

Most keen cacti growers generally dislike the use of grafting for cacti. The resultant plants are often referred to as "lollipops" due to the appearance of the grafted plant on a stick. Grafted cacti also tend to be frowned upon in competitive shows and marked down. This is probably due to the belief that grafted plants are easily cultivated, grow faster than non-grafted plants, and often appear bloated and atypical. As long as the scion is not lacking chlorophyll, growers will often plant the graft lower in the pot and hide the rootstock with a gravel top dressing.

Case study: *Echinocereus* on *Trichocereus*

The rootstock is selected to match the scion material required to be grafted. In this example, *Trichocereus spachianus* has been selected to form the rootstock. The rootstock should be actively growing, have plenty of fresh growth, and the portion of plant tissue where the graft will be made is not more than 12 months old.

A sharp, sterile blade such as a scalpel is used to cut off the top of the rootstock, slightly higher than the point at which the graft will be made. The top portion can be left to dry for a week or two then stood in a pot of dry compost to be rooted, grown on, and used for grafting the following year.

The top of the rootstock is then cleaned up, making a horizontal cut surface that is free from damage or bumpy tissue. The horizontal cut is important in allowing the elastic bands to hold the scion in place. If the cut surface of the stock is at an angle, there is a danger that the tension from the elastic bands will slide the scion off the stock.

Once the rootstock cut has been cleaned up, carefully bevel cut the ribs around the surface, resulting in a slightly raised flat surface on which to place the scion. The reason for this bevel cut is that as the cut surfaces dry out, the soft inner tissue will shrink down a little, whereas the tougher epidermal tissue will remain more rigid and pucker slightly, possibly pushing the scion off the join. Bevel cutting allows the cut surface to callous over without puckering.

After cutting the rootstock, the scalpel is disinfected before removing the offset that will be used as the scion. In this example, an offset of *Echinocereus inermis* is being grafted since the parent plant is starting to decline significantly. Once it has been cut from the parent plant, the cut is cleaned up, removing any old or damaged tissue, creating a nice clean flat surface.

Old plant of *Echinocereus inermis* with small offset that will be used as the scion.

The scion is slid onto the cut surface of the rootstock. Both surfaces should be quite juicy so that the scion will move easily. Slight pressure is applied to the scion as it is slid around the surface of the stock to flush any air bubbles out of the joint. Then the scion is aligned to give the best possible overlap of vascular material and kept steady.

While gently holding the scion in place, stretch an elastic band over it. It is essential that the mid point of the elastic band is placed over the centre of the scion so that the tension is equal on both sides, otherwise the scion will slide off. Choose elastic bands of a suitable size that will fit snugly around the pot and up to the cut surface of the rootstock. This ensures that when the scion is placed on the rootstock it will be held in place gently, but firmly, allowing the two cut surfaces

time to knit together. Too loose and the scion may not adhere, pushing itself off as it dries out. Too tight and the scion may be pulled off to one side or damaged by the elastic biting into the scion tissue.

If the first elastic band holds the scion steady, the second elastic band can be attached at right angles to the first. Again, ensure that the elastic tension is equal at both sides where the band touches the scion.

For small grafts of fresh tissue, two medium-sized bands are sufficient to hold the scion in place. However, if grafting larger scions or slightly older tissue (which is not ideal), it might be necessary to add more elastic bands to improve contact.

If the growing point of the scion is particularly delicate or spiny, it might be beneficial to use a little padding to cushion the area in contact with

Well-rooted cutting of *Trichocereus spachianus* to be used as the rootstock.

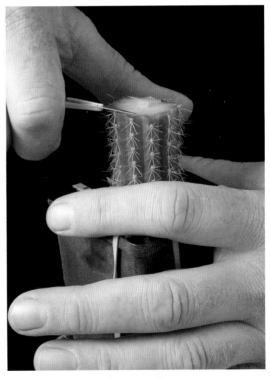

Cleaning up the top of the rootstock to provide a smooth, horizontal surface.

Completion of bevelled cut

Final preparation of the rootstock prior to attaching the scion. This shows the juicy cut surface ready to accept a prepared scion. The vascular cylinder can be observed at the centre of the cut.

Cutting the offset from the parent plant.

Rubber ties holding the completed graft.

the elastic bands. A small square of folded kitchen paper works well.

The grafted plant is then carefully placed in a warm, slightly humid environment away from direct sunlight, ideally in a suitable propagator. Given that both stock and scion were active at the time of grafting and that the grafting took place during the summer months, the elastic bands can usually be removed after two weeks and the graft moved to a position of better light. If grafting larger scions and using more than two elastic bands, the bands could gradually be removed after two weeks, to reduce the pressure on the scion. If all looks well with the union the rest can be removed a week or two later.

After the elastic bands are removed, the union will still be weak for a while, so care must be taken not to knock or bang the plant for the rest of the season at least, after which, good healthy grafts will be quite robust.

If the graft has taken properly, new growth in the scion will be apparent within a couple of weeks. If the scion seems firmly attached, but there is no visible sign of growth by the next growing season, it is highly probable that there was not a good union of vascular material. Eventually the scion will dry up and be detached easily.

A variegated form of *Lobivia* one year after grafting.

9 The Future of Grafting

AS A METHOD OF PROPAGATION, GRAFTING will possibly decline in the future as techniques in rooting cuttings and micropropagation continue to develop. Shrub roses in mainland Europe are more often propagated from cuttings now than by budding. Species like *Acer palmatum* can be rooted successfully even though many propagators still prefer to use grafting. Many ornamental trees, like *Prunus serrulata* 'Kanzan', can also be rooted by cuttings, although this is still not commonly used.

There are definite advantages to growers of plants propagated from cuttings rather than from grafting. In a survey of growers on the future of grafting, Marc Legare of the Quebec Institute for the Development of Ornamental Horticulture identified a shortage of skilled grafters as being a problem for nurseries. The move to cutting propagation is unlikely to reduce this problem. In addition, plants are more expensive to produce from grafts than from cuttings. Cuttings also avoid the problems of incompatibility or suckering encountered in grafts. Even the use of micropropagation is becoming more viable for relatively small batches of cuttings as culture techniques and aftercare develop.

The use of grafting extends beyond being simply a method of propagation, however, particularly in pest and disease resistance and control of vigour. Grape vines will continue to be grafted on a large scale, as there seems to be no other solution to the phylloxera problem. Grafting fruit and nut trees will continue to be used for the benefit of vigour control by the selected rootstock. The development of cherry production under protection was only possible with the development of dwarfing rootstocks like the Gisela series. This has rejuvenated cherry production in the United Kingdom and even allowed it to start in the north of Scotland.

Perhaps there is scope for other fruit crops to be grown under protection in the United Kingdom. For example, research into dwarf rootstocks for peaches and nectarines is being undertaken. The development of vegetatively propagated rootstocks for the likes of oaks would be very valuable. Rootstocks of known cambial peroxidase enzymes might enable growers to overcome the incompatibility issues in these species. Alternatively, the development of a rapid, inexpensive test for peroxidase type could also help to make oak cultivars more available.

One area of grafting that does seem to be set to expand is in the use of grafted plants for vegetable salad crops. Glasshouse production of tomatoes and aubergines is almost exclusively from grafted plants and sweet peppers and cucumbers are becoming more available. The numbers of plants required, and the uniformity of seedlings prior to grafting, has made the development of mechanized grafting possible and economic.

Young grafted vegetable plants are also be-

coming more available to the amateur grower and this trend will probably lead to their increasing use by gardeners, unlike in the 1960s when only seed was available and gardeners had to graft their own plants. In late 2013, the TomTato was released to the gardening markets in New Zealand and the United Kingdom. The plant is a graft combination of a tomato scion onto a potato rootstock. Although known since the nineteenth century, up until now this combination has just been a novelty. The development of small potato tubers that produce stems of the correct diameter to match the seedling tomatoes has allowed the grafting of these plants to be carried out economically.

Grafting of other crops like fruit trees is less likely to be mechanized. Legare identified a number of reasons for this: the rootstock and scion are more variable in size with fruit trees making matching the cuts by machine difficult; the blades will become blunt more quickly than with soft vegetable grafts, and where chip buds have been taken by machine the buds are often felt to be too small. There are aids to grafting, equipment that will prepare the cuts, but often a good grafter is faster than this equipment.

The great interest and enthusiasm of professional growers for the subject of grafting is obvious. Improving the success rate of graft take is very important to them, as is the quality of the plants produced. All of those I interviewed for this book mentioned possible changes they were considering making to their production.

In 2008 and 2012, the Great Britain team dominated cycling events at the Olympics. A major reason for its domination of the gold medals was Matt Parker, head of marginal improvement. He headed a group that looked at the whole process of cycle racing and came up with small improvements that would take the tenths of seconds off times that make the difference between gold and silver. There was the use of a small amount of alcohol rubbed around the tyres before the start of a race. This removed dust and gave better grip, giving the cyclists a slightly faster start. There was also the development of "hot pants," heated leggings used between warm up and the start of the race so that muscles would not cool down. In themselves, each made only a little difference, but together these improvements helped good riders take fractions of a second off their times and become great riders.

This is what the growers are doing. They know where improvements could be made and are taking ideas from researchers, societies like the International Plant Propagators' Society, and other growers to make the marginal gains in their plant

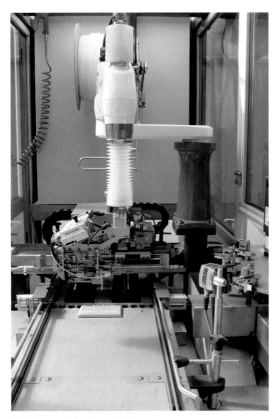

Grafting machine for salad crops.

production, reduce losses and improve efficiency and quality. This might be by using fans in a tunnel to circulate the air while maintaining humidity and thereby reduce fungal problems. It might be by grafting with single bud scions so that once growth started all the plant energy goes straight away into the growth of the one bud. It might be by using fog in place of polythene tents to control humidity more precisely. Perhaps grafting in late summer rather than early spring will improve the first year's growth. Perhaps all the grafting can be done in early spring so other nursery tasks can be carried out more efficiently in the summer. The best growers are never fully satisfied with their results and are always looking for the marginal improvement.

The problem with marginal improvements, however, is that there is little innovation. This can be a problem in research where reputations and funding often depend on publishing peer-reviewed papers and there is a temptation to do safe research for marginal improvements. In 2007, one of the winners of the Nobel Prize for Medicine was Italian American Mario Capecchi for his work on the mouse genome that led the way for all gene therapy. Mario had worked under James Watson of DNA fame, at Harvard University, but left to set up his own Molecular Biology Department at Utah University because Harvard had become "a bastion of short-term intellectual gratification." There were too many marginal improvements but little innovation. Capecchi found funding hard to come by, and probably quite rightly, because what he proposed (in knocking a gene out of the mouse genome) had never been done and few thought it possible at the time. Innovation is risky and for every Mario Capecchi there will be ten or more whose ideas come to nothing. But his success led to a medical breakthrough that could never occur with marginal gains.

Growers and researchers have made many marginal improvements in grafting in recent years and these have been, and continue to be, very valuable. However, there seems to have been little real innovation in grafting since the 1970s. Around this time, chip budding was adopted by most tree growers following research led by Brian Howard at East Malling Research. The hot-pipe system, first developed by Harry Lagerstedt at Oregon State University, was being taken up by growers to improve their bench grafting results. Micrografting was also first being used to help produce virus-free plant material. In the survey by Legare, all the people questioned agreed that there would not be a spectacular development in grafting. But perhaps there is a Mario Capecchi of the grafting world out there who can come up with the next innovative idea for grafting. If it is you, good luck with obtaining the funding to develop your idea.

In the meantime, if you are grafting plants for a living or as a hobby, continue to question and improve what you are doing. There are always improvements to be made as our understanding of plant growth develops and equipment and materials improve. This is what makes plant propagation so interesting. You will never know everything, and there will always be someone from whom you can learn. The motto of the International Plant Propagators' Society is "Seek and Share." Don't just look at what you are doing; share your knowledge with others and you will assuredly learn more from meeting other propagators that will help you become a better grafter.

APPENDIX

Grafting Methods and Rootstocks for Selected Species

The following conventions are used in the three tables that follow.

Scion normally refers to cultivars of the species listed in this column.

Rootstock refers to all the plants that are recorded as being successful rootstocks with the named scions. Where several species are named, further research should be carried out to identify the one(s) most suitable for a particular situation.

Grafting method refers to one or more of the seven techniques discussed in this book and described here in brief.

1. **Apical, bareroot spring**. Apical graft (any graft where the rootstock is cut back to the required height before the scion is attached using dormant bareroot rootstock, cold callused in early spring. After grafting, plants are potted and stood in cold greenhouse or planted out into field.

2. **Apical, pot-grown, spring**. Apical graft using pot-grown rootstock, cold callused in early spring. Rootstock should have at least 6.5 mm fresh white roots showing. After grafting, plants maintain at four degrees above outside temperature and ventilate at 10°C.

3. **Hot-pipe**. Apical graft using bareroot or pot-grown rootstocks, placed on hot callus pipe after grafting in early spring. Temperatures provided are a guide for the hot-pipe. If no temperature is given, 18°C is recommended. The temperature can then be adjusted until the best results are achieved. Care should be taken if using temperatures higher than 18°C. It is important that the temperature of the pipe does not fluctuate above the highest temperature given.

4. **Apical, pot-grown, summer**. Apical graft using pot-grown rootstocks in late summer. Aftercare involves control of humidity under contact polythene and careful management of light levels and temperature.

5. **Side, summer**. Side graft (any graft where the top of the rootstock remains attached until the union has formed at which point the rootstock is headed back) using pot-grown rootstocks in late summer. Aftercare involves control of humidity under contact polythene and careful management of light levels and temperature.

6. **Side, spring**. Side graft using pot-grown rootstocks in early spring. Normally aftercare involves maintaining humidity under a polythene tent and a base temperature of 18–20°C and an air temperature of 15–18°C. Cooler temperatures below 15°C are becoming more popular; the graft union is slower to form, but there are fewer *Botrytis* problems.

7. **Budding**. Budding onto field-grown rootstocks in summer. Chip budding will normally be used unless otherwise stated.

Comments include additional information that may assist in decision making about how to graft a specific plant. Rootstocks listed in this column may be suitable for particular applications.

Scion-rootstock combinations

SCION	ROOTSTOCK	GRAFTING METHOD							COMMENTS
		Apical, bare-root, spring	Apical, pot-grown, spring	Hot-pipe (temperature)	Apical, pot-grown, summer	Side, summer	Side, spring	Budding	
Abies alba	*Abies alba*						●		*Abies firma* may be a suitable rootstock for low-oxygen and high soil-temperature conditions
	A. concolor						●		
	A. balsamea						●		
	A. procera						●		
A. concolor	*A. concolor*						●		
A. homolepis	*A. firma*						●		
Acer japonicum	*Acer palmatum*	●	●			●	●		*Acer palmatum* cultivars can also be budded in areas with reliable, warm summers.
A. palmatum		●	●			●	●		Summer grafting is carried out around July as *Acer* only has one flush of growth a year. Apical, pot-grown graft has been used successfully as long as aftercare of air humidity, temperature, light, and water management are strictly applied.
Acer—other species	Use the same species as cultivars to be propagated apart from the exceptions listed here			● (16–20°C)	●	●		●	
A. rubrum	*A. rubrum*			● (16–20°C)					Delayed compatibility can occur with *A. rubrum* cultivars although they are grafted onto seedlings of *A. rubrum*.
A. griseum	*A. nikoense*			● (16–20°C)	●	●		●	In general, when selecting rootstock species for rare acers, follow the taxonomic series classification, or certainly within the section.
A. heldreichi				● (16–20°C)	●	●		●	
A. trauvetteri				● (16–20°C)	●	●		●	There are also "milky" sap groups and "non-milky" groups. Graft within that character; for example, *A. platanoides* (milky) would be the choice for *A. cappadocicum, A. catalpifolium,* and *A. lobelii* cultivars. These are not compatible on *A. pseudoplatanus,* a non-milky example.
A. velutinum	*A. pseudoplatanus*			● (16–20°C)	●	●		●	
A. ×freemanii	*A. rubrum*			● (16–20°C)	●	●		●	
A. truncatum	*A. platanoides*			● (16–20°C)	●	●		●	*Acer rubrum* for *A. pentaphylla* scion and *A. saccharum* for *A. griseum* scion may be a suitable rootstock for low-oxygen and high soil-temperature conditions.

SCION	ROOTSTOCK	GRAFTING METHOD							COMMENTS
		Apical, bare-root, spring	Apical, pot-grown, spring	Hot-pipe (temperature)	Apical, pot-grown, summer	Side, summer	Side, spring	Budding	
Aesculus ×carnea	Aesculus hippocastanum			● (27°C)				●	Grafts should be removed from the hot-pipe after 21 days. Whip-and-tongue preferred to a whip graft as slow-to-form graft union and movement at the union.
	A. ×carnea			● (27°C)				●	
A. arguta	A. hippocastanum			● (27°C)				●	
A. glabra				● (27°C)				●	Best to select small buds that can be covered when tying-in. If large bud is selected, leave the bud exposed when tying-in to avoid damage.
A. hippocastanum				● (27°C)				●	
A. sylvatica				● (27°C)				●	
A. indica	A. hippocastanum			● (27°C)				●	Preferably select the type species as rootstock as A. hippocastanum can result in stem overgrowth at union.
	A. indica			● (27°C)				●	
Alnus ×cordinca	Alnus cordata		●				●		Apical, pot-grown graft has been used successfully as long as aftercare of air humidity, temperature, light, and water management are strictly applied.
A. glutinosa	A. glutinosa		●				●		
Amelanchier alnifolia	Amelanchier canadensis		●	● (20°C)			●	●	
A. canadensis	A. lamarckii		●	● (20°C)			●	●	Deep-planted bareroot bench grafts sucker less than field-budded plants.
A. lamarckii	Sorbus intermedia		●	● (20°C)			●	●	
A. laevis			●	● (20°C)			●	●	
Aralia chinensis	Aralia chinensis						●	●	
Arbutus ×andrachnoides	Arbutus unedo					●	●		
Berberis linearifolia	Berberis thunbergii		●	●					Berberis thunbergii is used for difficult-to-root evergreen species and cultivars. Berberis thunbergii 'Atropurpurea' is sometimes used as the rootstock as suckers will have the distinctive red foliage.
B. ×lologensis			●	●					

Scion-rootstock combinations, *continued*

SCION	ROOTSTOCK	Apical, bare-root, spring	Apical, pot-grown, spring	Hot-pipe (temperature)	Apical, pot-grown, summer	Side, summer	Side, spring	Budding	COMMENTS
Betula ermanii	*Betula pendula*		●	● (16–23°C)			●		Apical graft becoming more popular when used with a callus hot-pipe. Grafts should be removed from hot-pipe after 18 days.
B. grossa			●	● (16–23°C)			●		
B. korshinskyi			●	● (16–23°C)			●		
B. pendula	*B. pendula*		●	● (16–23°C)			●		Side graft may be preferred due to early and profuse sap rise.
B. ×caerulea	*B. populifolia*		●	● (16–23°C)			●		Chip budding may be used if reliable warm summers
	B. papyrifera		●	● (16–23°C)			●		*Betula pubescens* may be useful for wet soils.
B. ×intermedia	*B. platyphylla*		●	● (16–23°C)			●		*Betula nigra* may be a suitable rootstock for low-oxygen and high soil-temperature conditions.
Calocedrus decurrens	*Calocedrus decurrens*					●			
	Thuja occidentalis					●			
Camellia reticulata	*Camellia japonica*	●					●		Mainly propagated by cuttings but some cultivars ('Pink Pagoda') are poor rooters and have weak roots. Use strong-growing cultivars like 'Debutante' as rootstock.
Carpinus betulus	*Carpinus betulus*			● (22°C)	●	●	●		Winter graft preferred.
C. caroliniana				● (22°C)	●	●	●		Grafts should be removed from the hot-pipe after 16 days.
C. japonica				● (22°C)	●	●	●		Large bareroot seedlings with thick scions produce larger plants more quickly than thin pot-grown rootstocks.
C. turczaninowii				● (22°C)	●	●	●		Strict aftercare of air humidity, temperature, light, and water management are required with apical graft.
Carya illinoensis	*Carya illinoensis*							●	Selected cultivars can be used as rootstocks to provide more fibrous root systems.
Castanea sativa	*Castanea sativa*			● (22°C)					Have been successfully chip budded in the field in Australia.
Catalpa bignonioides	*Catalpa bignonioides*	●	●						May be top-worked but usually low-worked to prevent swollen graft at eye level.
C. fargesii	*C. speciosa*	●	●						At Royal Botanic Gardens, Edinburgh, root grafts are used in early spring.

SCION	ROOTSTOCK	GRAFTING METHOD							COMMENTS
		Apical, bare-root, spring	Apical, pot-grown, spring	Hot-pipe (temperature)	Apical, pot-grown, summer	Side, summer	Side, spring	Budding	
Cedrus species and cultivars	Cedrus deodara					●	●		Summer preferred
Celtis laevigata	Celtis laevigata						●		
C. occidentali	C. occidentali						●		
Cercis canadensis	Cercis canadensis		●	●			●		
C. chinensis			●	●			●		Cercis chinensis may be a suitable rootstock for low-oxygen and high soil-temperature conditions as well as C. canadensis.
C. griffithii			●	●			●		
C. occidentalis			●	●			●		
C. siliquastrum	C. siliquastrum		●	●			●		
Chamaecyparis lawsoniana	Chamaecyparis lawsoniana					●	●		
C. obtusa	C. lawsoniana					●	●		
	C. pisifera					●	●		Chamaecyparis pisifera or C. thyoides may be a suitable rootstock for low-oxygen and high soil-temperature conditions.
	Thuja occidentalis					●	●		
C. nootkatensis	C. lawsoniana					●	●		
	Platycladus orientalis					●	●		
Chionanthus virginicus	Fraxinus excelsior	●		●					Rootstocks usually grown from seed.
Cladrastis lutea	Cladrastis lutea							●	
Clematis species and cultivars	Clematis vitalba	●		●					Nurse graft on roots that is then potted deeply to encourage scions to root. Largely superseded by cutting propagation.

Scion-rootstock combinations, *continued*

SCION	ROOTSTOCK	Apical, bare-root, spring	Apical, pot-grown, spring	Hot-pipe (temperature)	Apical, pot-grown, summer	Side, summer	Side, spring	Budding	COMMENTS
Cornus controversa	*Cornus controversa*		●	● (22°C)	●	●	●	●	Apical graft has been used successfully as long as aftercare of air humidity, temperature, light, and water management are strictly applied.
	C. alternifolia		●	● (22°C)	●	●	●	●	
C. nuttallii	*C. florida*		●	● (22°C)	●	●	●	●	
	C. nuttallii		●	● (22°C)	●	●	●	●	
C. florida	*C. florida*		●	● (22°C)	●	●	●	●	
C. kousa	*C. kousa* var. *chinensis*		●	● (22°C)	●	●	●	●	
C. 'Eddie's White Wonder'			●	● (22°C)	●	●	●	●	T-budding in late summer and early autumn may be better than chip budding with *C.* 'Eddie's White Wonder'.
Corylus avellana	*Corylus avellana*		●	● (22°C)	●				The grafts should be removed from the hot-pipe after 14 days.
	C. colurna		●	● (22°C)	●				*Corylus colurna* has advantage of easier to distinguish sucker growth.
	C. maxima		●	● (22°C)	●				Deep planting of bare root grafted plants sucker less than pot-grown rootstocks.
C. heterophylla	*C. colurna*		●	● (22°C)	●				Summer grafting can give greater maiden growth by end of first season.
Cotoneaster ×*watereri*	*Cotoneaster bullatus*							●	*Cotoneaster bullatus* prone to suckering.
Crataegus species and cultivars	*Crataegus monogyna*	●	●	●				●	*Crataegus aestivalis* may be a suitable rootstock for low-oxygen and high soil-temperature conditions. *Crataegus phaenopyrum* is an alternative rootstock for many *Crataegus* species.
Cryptomeria japonica	*Cryptomeria japonica*						●		Seedlings or rooted cuttings can be used for the rootstock.
Cupressus glabra	*Cupressus macrocarpa*					●			*Cupressus bakeri* or *C. arizonica* may be a suitable rootstock for low-oxygen and high soil-temperature conditions for *C. sempervirens*.
C. macrocarpa						●			
C. sempervirens	*C. macrocarpa*					●			
	C. sempervirens					●			
Cytisus battandieri	*Laburnum anagyroides*		●	●					

SCION	ROOTSTOCK	GRAFTING METHOD							COMMENTS
		Apical, bare-root, spring	Apical, pot-grown, spring	Hot-pipe (temperature)	Apical, pot-grown, summer	Side, summer	Side, spring	Budding	
Daphne species	Daphne mezereum		●		●				Growth may be best on D. mezereum using root sections or seedlings.
	D. laureola		●		●				Grafting is effective for many of the slow-growing species.
Davidia involucrata	Davidia involucrata	●		●	●				
Elaeagnus macrophylla	Elaeagnus pungens				●	●			
Euonymus europaeus	Euonymus europaeus	●		● (18-20°C)					
Fagus engleriana	Fagus orientalis			● (24°C)	●		●		Grafts should be removed from the hot-pipe after 21 days.
F. orientalis	F. sylvatica			● (24°C)	●		●		Can be grafted in the summer but more usually winter grafted.
F. sylvatica				● (24°C)	●		●		Fagus grandiflora may be a suitable rootstock for low-oxygen and high soil-temperature conditions.
Fraxinus americana	Fraxinus americana			●				●	
F. angustifolia	F. excelsior			●				●	Most Fraxinus are open-ground budded.
F. excelsior	F. pennsylvanica var. lanceolata			●				●	
F. pennsylvanica				●				●	
F. bungeana	F. ornus			●				●	
F. floribunda				●				●	Overgrowths at the graft union are likely to occur when the smaller species of Fraxinus are worked onto vigorous rootstocks.
F. mariesii				●				●	
Ginkgo biloba	Ginkgo biloba			● (22°C)				●	
Gleditsia triacanthos	Gleditsia triacanthos var. inermis	●		● (20°C)				●	Gleditsia triacanthos is used but var. inermis is a thornless rootstock that makes handling easier.
Halesia species and cultivars	Halesia carolina	●	●						
H. diptera	H. monticola	●	●						

Scion-rootstock combinations, *continued*

SCION	ROOTSTOCK	GRAFTING METHOD							COMMENTS
		Apical, bare-root, spring	Apical, pot-grown, spring	Hot-pipe (temperature)	Apical, pot-grown, summer	Side, summer	Side, spring	Budding	
Hamamelis ×*intermedia*	*Hamamelis virginiana*		●	●	●	●			*Hamamelis vernalis* may produce excessive suckering but in Ireland it has been shown that a selected clone, giving little suckering, can be propagated from cuttings.
H. japonica	*H. japonica*		●	●	●	●			
H. mollis	*H. vernalis*		●	●	●	●			Chip-budded in the field in New Zealand. Could be used in UK if bud-wood is forced under protection to get ripe buds by the second week in August.
Hibiscus syriacus	*Hibiscus syriacus*			●					2-year-old seed-raised hibiscus rootstocks used. Successfully root from hardwood cuttings but subsequent growth is considerably slower.
Ilex aquifolium	*Ilex aquifolium*	●						●	Budding may be done in mid to late spring also. High working is sometimes carried out. *Ilex* 'Nellie Stevens' may be a suitable rootstock for low-oxygen and high soil-temperature conditions. *Ilex cornuta* 'Burfordii' can be used to reduce root rot problems.
Juglans regia	*Juglans regia*			● (24–32°C)					Grafts should be removed from the hot-pipe after 21 days. Whip-and-tongue graft is not recommended for *Juglans* because of its pithy nature.
	J. nigra			● (24–32°C)					
J. nigra	*J. nigra*			● (24–32°C)					
Juniperus chinensis	*Juniperus chinensis* 'Hetzii'					●	●		Most cultivars propagated by cuttings.
J. communis						●	●		
J. virginiana						●	●		
J. scopulorum	*Juniperus chinensis* 'Hetzii'					●	●		
	J. chinensis 'Glauca Hetzii'					●	●		
	J. virginiana					●	●		
Kalmia latifolia	*Kalmia latifolia*		●				●		
+*Laburnocytisus* 'Adamii'	*Laburnum anagyroides*					●	●		

SCION	ROOTSTOCK	GRAFTING METHOD							COMMENTS
		Apical, bare-root, spring	Apical, pot-grown, spring	Hot-pipe (temperature)	Apical, pot-grown, summer	Side, summer	Side, spring	Budding	
Laburnum anagyroides	*Laburnum anagyroides*	●	●	●				●	
L. ×watereri 'Vossii'	*L. alpinum*	●	●	●				●	
	L. vulgare	●	●	●				●	
Larix decidua	*Larix decidua*		●				●		
	L. kaempferi		●				●		
L. sibirica	*L. leptolepis*		●				●		
Ligustrum lucidum	*Ligustrum ovalifolium*		●				●		*Ligustrum vulgare* is the hardier rootstock.
	L. vulgare		●				●		
Liquidambar styraciflua	*Liquidambar styraciflua*		●	●					Many cultivars successfully root from softwood cuttings.
Liriodendron chinense	*Liriodendron tulipifera*			● (24°C)					
L. tulipifera				● (24°C)					
Magnolia campbellii	*Magnolia campbellii*		●	● (24°C)					Chip budding of pot-grown rootstocks can also be carried out in the summer.
M. dawsoniana	*M. dawsoniana*		●	● (24°C)					In subgenus *Magnolia*, section *Magnolia*, *M. ×wiesneri* (syn. *M. ×watsonii*) can be grafted onto seedlings of *M. hypoleuca*.
M. sargentiana	*M. kobus*		●	● (24°C)					In subgenus *Yulania*, section *Yulania*, *M. cylindrical*, *M.* 'Albatross', *M.* 'Yellow Bird', *M. campbellii*, *M. dawsoniana*, and *M. sprengeri* can be grafted onto rootstock of *M.* 'Heaven Scent' which is propagated from cuttings.
M. sprengeri	*M. sprengeri*		●	● (24°C)					
M. grandiflora	*M. grandiflora*		●	● (24°C)					
Other species	*M. kobus*		●	● (24°C)					*Magnolia virginiana* may be a suitable rootstock for low-oxygen and high soil-temperature conditions for scions of *M. sieboldii* and *M. wilsonii*.
	M. ×soulangeana		●	● (24°C)					

Scion-rootstock combinations, *continued*

SCION	ROOTSTOCK	Apical, bare-root, spring	Apical, pot-grown, spring	Hot-pipe (temperature)	Apical, pot-grown, summer	Side, summer	Side, spring	Budding	COMMENTS
Malus species and cultivars Ornamental crab apples	MM106	●						●	MM106 is commonly used for standard trees. M27 is used for patio plants. *Malus* 'Antonovka' and *M.* 'Columba' are used in areas of very cold temperatures. Hot pipe can be used for faster graft union.
	MM111	●						●	
	M25	●						●	
	M27	●						●	
	M. baccata	●						●	
	M. sylvestris	●						●	
	M. 'Antonovka'	●						●	
	M. 'Columba'	●						●	
	M. 'Bittenfelder'	●						●	
Mespilus germanica	*Quince A*	●		●					
	Crataegus monogyna	●		●					
Metasequoia glyptostroboides	*Metasequoia glyptostroboides*		●						
Morus alba	*Morus alba*			● (20°C)					A side graft can be used.
M. bombycis	*M. alba* var. *tatarica*			● (20°C)					
M. latifolia	*M. nigra*			● (20°C)					
Nyssa sylvatica	*Nyssa sylvatica*		●	●					
Paeonia lutea	*Paeonia lactiflora*				●				Plant deep to encourage scion rooting. Can be chip budded on to rootstock root pieces. Place bare roots into a moist media under contact polythene at 20–24°C. After graft takes, cold store and plant in spring.
P. delavayi	*P. officinalis*				●				
P. potaninii					●				
P. suffruticosa	*P. lactiflora*				●				
Parrotia persica	*Hamamelis virginiana*		●	●	●				
Photinia ×fraseri	*Chaenomeles japonica*		●	●					When potting on, plant deep to encourage self-rooting.
	Crataegus monogyna		●	●					

SCION	ROOTSTOCK	GRAFTING METHOD							COMMENTS
		Apical, bare-root, spring	Apical, pot-grown, spring	Hot-pipe (temperature)	Apical, pot-grown, summer	Side, summer	Side, spring	Budding	
Picea abies	*Picea abies*					●	●		*Picea chinensis* is preferred rootstock.
P. breweriana	*P. chinensis*					●	●		The union forming before winter enables improved scion growth in the spring.
P. omorika						●	●		*Picea omorika* or *P. orientalis* may be a suitable rootstock for low-oxygen and high soil-temperature conditions for *P. breweriana*.
P. orientalis						●	●		
P. pungens						●	●		
Pinus species 2-needled examples:									Normally, graft 2-needled pines on 2 needled rootstock and similarly "3 on 3" and "5 on 5."
P. densiflora	*P. contorta*					●	●		Some propagators have found *P. contorta* to be compatible with 2-, 3-, and 5-needled pines.
P. mugo	*P. sylvestris*					●	●		*Pinus sylvestris* does not give best root system and is prone to root rots. *P. mugo* gives a bushy dense root system but has very branched and knotty angular growth, making it difficult to find a straight stem on which to graft.
P. sylvestris	*P. mugo*					●	●		
	P. uncinata					●	●		
5-needled examples:									
P. parviflora	*P. strobus*					●	●		*Pinus uncinata*, which is closely related to *P. mugo* and has a similar root system, produces a straighter stem and is easier to graft onto.
P. pumila						●	●		
P. strobus						●	●		
Pittosporum eugenioides	*Pittosporum tenuifolium*						●		
P. tenuifolium							●		
Platanus ×hispanica	*Platanus ×hispanica*	●							
Platycladus orientalis	*Platycladus orientalis*					●			Many *P. orientalis* cultivars root successfully from cuttings.
Populus species	*P. canadensis*	●		●					
	P. alba	●		●					Can graft onto unrooted leafless winter cuttings.
	P. ×candicans	●		●					
P. tremula	*P.* Brooks No. 6	●		●					

Scion-rootstock combinations, *continued*

SCION	ROOTSTOCK	GRAFTING METHOD							COMMENTS
		Apical, bare-root, spring	Apical, pot-grown, spring	Hot-pipe (temperature)	Apical, pot-grown, summer	Side, summer	Side, spring	Budding	
Prunus dulcis	Brompton	●	●	● (20°C)				●	
	St. Julien A	●	●	● (20°C)				●	
	Prunus dulcis	●	●	● (20°C)				●	
	P. persica	●	●	● (20°C)				●	
P. armeniaca	Brompton	●	●	● (20°C)				●	
	St. Julien A	●	●	● (20°C)				●	
	P. armeniaca	●	●	● (20°C)				●	
	P. persica	●	●	● (20°C)				●	
P. padus	*P. padus*	●	●	● (20°C)				●	
P. ×*schmittii*	*P. avium*	●	●	● (20°C)				●	Thin-barked species like *P. subhirtella* are difficult to bud but graft easily.
P. ×*yedoensis*	Mazzard F12/1	●	●	● (20°C)				●	St Julien A will produce a smaller tree than other rootstocks.
P. ×*hillieri*	Colt	●	●	● (20°C)				●	*Prunus padus* prone to suckering.
P. sargentii		●	●	● (20°C)				●	Colt may not be hardy in very cold climates.
P. serrula		●	●	● (20°C)				●	
P. subhirtella		●	●	● (20°C)				●	
P. serrulata		●	●	● (20°C)				●	
P. ×*incam* 'Okamé'		●	●	● (20°C)				●	
P. virginiana	*P. padus*	●	●	● (20°C)				●	
	P. virginiana	●	●	● (20°C)				●	
P. persica	*P. persica*	●	●	● (20°C)				●	
	Brompton	●	●	● (20°C)				●	
	St. Julien A	●	●	● (20°C)				●	

SCION	ROOTSTOCK	GRAFTING METHOD							COMMENTS
		Apical, bare-root, spring	Apical, pot-grown, spring	Hot-pipe (temperature)	Apical, pot-grown, summer	Side, summer	Side, spring	Budding	
P. cerasifera	*P. cerasifera*	●	●	● (20°C)				●	
	Myrobalan B	●	●	● (20°C)				●	
	St. Julien A	●	●	● (20°C)				●	
P. ×*amygdalo-persica*	*P. persica*	●	●	● (20°C)				●	
	St. Julien A	●	●	● (20°C)				●	
P. triloba	*P. cerasifera*	●	●	● (20°C)				●	
	Brompton	●	●	● (20°C)				●	
	Myrobalan B	●	●	● (20°C)				●	
	St. Julien A	●	●	● (20°C)				●	
P. mume	*P. cerasifera*	●	●	● (20°C)				●	
	St. Julien A	●	●	● (20°C)				●	
	P. persica	●	●	● (20°C)				●	
Pyrus calleryana	*Pyrus calleryana*	●	●	●				●	*Pyrus communis* may give some incompatibility problems and be susceptible to fireblight.
P. salicifolia	*P. communis*	●	●	●				●	
	P. ussuriensis	●	●	●				●	Pot-grown rootstocks will give more even maiden growth than bare root.

Scion-rootstock combinations, *continued*

SCION	ROOTSTOCK	GRAFTING METHOD							COMMENTS
		Apical, bare-root, spring	Apical, pot-grown, spring	Hot-pipe (temperature)	Apical, pot-grown, summer	Side, summer	Side, spring	Budding	
Quercus species and cultivars									Grafts should be removed from the hot-pipe after 17 days.
Q. castaneifolia	*Q. cerris*			● (24°C)	●	●	●		It is important to select the rootstocks from within the appropriate group for long-lasting unions. For example, in subgenus *Quercus*, both section *Quercus* (the white oaks) and the closely related section *Mesobalanus* have considerable latitude to match within the subgenus. *Quercus robur* is the usual rootstock for the subgenus. *Quercus muehlenbergii* is a possibility for high-pH soils.
Q. coccinea	*Q. palustris*			● (24°C)	●	●	●		
	Q. rubra			● (24°C)	●	●	●		
Q. ×hispanica	*Q. ilex*			● (24°C)	●	●	●		
Q. frainetto	*Q. robur*			● (24°C)	●	●	●		
Q. macranthera	*Q. bicolor*			● (24°C)	●	●	●		Species in subgenus *Quercus*, section *Lobatae* (the red oaks), such as *Q. rubra*, have greater incompatibility problems, so it is important to match species. However, cultivars of *Q. rubra* grafted onto *Q. rubra* can give unpredictable results.
Q. petraea	*Q. michauxii*			● (24°C)	●	●	●		
Q. robur				● (24°C)	●	●	●		*Quercus virginiana* may be a suitable rootstock for low-oxygen and high soil-temperature conditions for U.S. West Coast and Mediterranean evergreen species.
Q. rubra	*Q. rubra*			● (24°C)	●	●	●		
Rhododendron species and cultivars	*Rhododendron* 'Cunningham's White'		●				●		Important to select rootstocks within same series as scion.
R. macabeanum	*R. grande*		●				●		Cutting-grafts of more difficult-to-root rhododendrons, using easy-to-root rootstocks (e.g., *R.* 'Catawbiense Boursault' or 'Catawbiense Grandiflorum'), have been successfully carried out in mid autumn.
R. sinogrande			●				●		
R. yakushimanum	*R.* 'Cunningham's White'		●				●		Saddle graft also used.
	R. Inkahro		●				●		*Rhododendron chapmanii* for small-leaved evergreen and *R.* atlanticum for deciduous types may be suitable rootstocks for low-oxygen and high soil-temperature conditions.
									Rhododendron yakushimanum cultivars graft well in late summer and early autumn
									Inkarho rhododendron rootstock is tolerant of high soil pH.
Robinia hispida	*Robinia pseudoacacia*	●	●	●				●	
R. ×margaretta		●	●	●				●	Normally top-worked.
R. pseudoacacia		●	●	●				●	

SCION	ROOTSTOCK	GRAFTING METHOD							COMMENTS
		Apical, bare-root, spring	Apical, pot-grown, spring	Hot-pipe (temperature)	Apical, pot-grown, summer	Side, summer	Side, spring	Budding	
Rosa cultivars	*Rosa canina* 'Inermis'							●	
	R. multiflora							●	T-bud once rind is slipping and scion buds are mature.
	R. laxa							●	*Rosa rugosa* is traditional rootstock for tree (standard) roses.
	R. rubiginosa							●	
	R. rugosa							●	
Salix species and cultivars	*Salix* ×*smithiana*	●							Great majority of *Salix* species root successfully from softwood and deciduous hardwood cuttings. See top-working chart for weeping cultivars.
	S. daphnoides	●							
Sequoia sempervirens	*Sequoia sempervirens*				●	●	●		A number of Californian clones root effectively from cuttings (e.g., 'Aptos Blue'). *Sequoiadendron giganteum* stated to be an appropriate rootstock and is hardier.
Sequoiadendron giganteum	*Sequoiadendron giganteum*					●	●		*Sequoia sempervirens* can be used but only in warmer climates due to hardiness.
Sorbus aucuparia	*Sorbus aucuparia*	●		●				●	
S. ×*arnoldiana*		●		●				●	
S. commixta		●		●				●	
S. hupehensis		●		●				●	
S. domestica	Quince A	●		●				●	
S. aria cultivars	*S. aria*	●		●				●	*Sorbus intermedia* generally preferred with *S. aria* cultivars as gives stronger growth after grafting. May present incompatibility issues with some cultivars ('John Mitchell' and 'Wilfrid Fox').
S. ×*hybrida*	*S. intermedia*	●		●				●	
S. megalocarpa	*S. meliosmifolia*	●		●				●	
	S. alnifolia	●		●				●	*Sorbus alnifolia* may be a suitable rootstock for low-oxygen and high soil-temperature conditions.

Scion-rootstock combinations, *continued*

SCION	ROOTSTOCK	GRAFTING METHOD							COMMENTS
		Apical, bare-root, spring	Apical, pot-grown, spring	Hot-pipe (temperature)	Apical, pot-grown, summer	Side, summer	Side, spring	Budding	
Syringa vulgaris	*Ligustrum ovalifolium*		●	●					*Syringa josikaea* and *S. reflexa* have very distinctive leaves, thus making easy-to-recognize suckers in gardens. *Syringa tomentella* results in fewer suckers.
	Syringa vulgaris		●	●					
	S. tomentella		●	●					*Ligustrum ovalifolium* is used as a nurse graft.
	S. josikaea		●	●					Softwood cutting propagation, open-ground budding, and layering also used.
	S. reflexa		●	●					Bench-grafted shrubs on nurse roots are superior to budded plants.
									Syringa oblata var. *dilatata* may be a suitable rootstock for low-oxygen and high soil-temperature conditions.
Taxodium distichum	*Taxodium distichum*	●							
Taxus baccata	*Taxus baccata*						●		Many *Taxus baccata* cultivars root successfully from cuttings.
Tilia amurensis	*Tilia cordata*	●	●	●				●	
T. mongolica	*T.* ×*europaea*	●	●	●				●	
T. cordata	*T. cordata*	●	●	●				●	*Tilia cordata* and *T. platyphyllos* can have incompatibility problems. East Malling Research has been working on vegetatively propagated clones of these rootstocks to overcome this problem; results are promising.
T. ×*euchlora*	*T. platyphyllos*	●	●	●				●	
T. americana	*T. americana*	●	●	●				●	
T. petiolaris	*T. platyphyllos*	●	●	●				●	
T. platyphyllos		●	●	●				●	
T. tomentosa		●	●	●				●	
Tsuga canadensis	*Tsuga canadensis*						●		Propagators have successfully used *T. canadensis* as rootstock for *T. heterophylla* clones and vice-versa.
T. heterophylla	*T. heterophylla*						●		
Ulmus glabra	*Ulmus glabra*	●	●	●				●	
U. ×*hollandica*	*U. pumila*	●	●	●				●	
U. procera	*U. procera*	●	●	●				●	

SCION	ROOTSTOCK	GRAFTING METHOD							COMMENTS
		Apical, bare-root, spring	Apical, pot-grown, spring	Hot-pipe (temperature)	Apical, pot-grown, summer	Side, summer	Side, spring	Budding	
Viburnum ×burkwoodii	*Viburnum dentatum*	●	●	●	●				
V. carlesii	*V. dilatatum*	●	●	●	●				
	V. lantana	●	●	●	●				
	V. opulus	●	●	●	●				Viburnums are grafted less often as techniques for rooting cuttings have developed. *Viburnum carlesii* can suffer large losses over winter from cuttings, and grafting is still used by some propagators.
V. ×bodnantense	*V. dentatum*	●	●	●	●				
V. ×juddii		●	●	●	●				
V. carlesii 'Compactum'	*V. setigerum*	●	●	●	●				
	V. ×rhytidophylloides 'Alleghany'	●	●	●	●				
V. lobophyllum	*V. ×rhytidophylloides* 'Willowwood'	●	●	●	●				
V. prunifolium	*V. ×pragense*	●	●	●	●				
V. rufidulum		●	●	●	●				
Wisteria floribunda	*Wisteria sinensis*			● (25°C)					
W. sinensis				● (25°C)					Can be cold callus grafted in mid spring using bareroot plants potted after grafting.
W. venusta				● (25°C)					
Zelkova serrata	*Zelkova serrata*						●	●	*Ulmus* species may also be suitable as a rootstock.

Top-worked tree combinations

SCION	ROOTSTOCK	GRAFTING METHOD							COMMENTS
		Apical, bare-root, spring	Apical, pot-grown, spring	Hot-pipe	Apical, pot-grown, summer	Side, summer	Side, spring	Budding	
Acer palmatum	Acer palmatum		●		●	●	●		Usually top-worked at a height of 20–60 cm. Stick-budding can be used for *A. palmatum*. The rootstock is prepared as for T-budding, but the scion has two pairs of buds and a sloping cut at the base. This is inserted behind the lifted bark of the rootstock.
A. platanoides	A. platanoides		●		●		●		Can be budded onto stems of *A. platanoides* 'Emerald Queen'. Use of an interstock can result in more upright growth of 'Globosum'.
A. pseudoplatanus	A. pseudoplatanus		●		●		●		
Aesculus ×carnea	Aesculus hippocastanum		●				●		More commonly open-ground low-worked by budding. *Aesculus hippocastanum* is likely to result in stem overgrowth at union.
A. pavia	A. hippocastanum		●				●		Preferable to use type species (e.g., *A. pavia*).
	A. pavia		●				●		
Aronia ×prunifolia	Sorbus aucuparia		●						
Betula nana	Betula pendula		●				●		Also open-ground budded in late summer. Often low-worked to prevent a swollen union at eye level. 'Trost's Dwarf' makes an effective patio tree.
B. pendula			●				●		
Caragana arborescens	Caragana arborescens		●				●		Whip-and-tongue is preferred by some growers for this species.
C. frutex			●				●		
C. tragacanthoides			●				●		
Cedrus atlantica	Cedrus deodara		●		●				Sometimes low-worked to avoid swollen union at eye level.
Cercidiphyllum. japonicum	Cercidiphyllum. japonicum	●	●						
Chamaecyparis obtusa	Chamaecyparis lawsoniana		●						*Chamaecyparis lawsoniana* should not be used in areas subject to root rot (*Phytophthora cinnamomi*).
	C. pisifera		●						

SCION	ROOTSTOCK	GRAFTING METHOD							COMMENTS
		Apical, bare-root, spring	Apical, pot-grown, spring	Hot-pipe	Apical, pot-grown, summer	Side, summer	Side, spring	Budding	
Corylus avellana	Corylus avellana	●	●						
C. maxima		●	●						
Cotoneaster adpressus	Cotoneaster bullatus	●	●						Cotoneaster frigidus often preferred as it has stronger stem.
C. apiculatus	C. frigidus	●	●						
C. dammeri	C. ×watereri cultivars	●	●						
C. ×hessei		●	●						Also open-ground grafted and budded.
C. horizontalis		●	●						
C. 'Hybridus Pendulus'		●	●						
Crataegus laevigata	Sorbus aucuparia	●							Double-working the already-budded rootstock should provide a good straight stem.
C. monogyna		●							
Cytisus battandieri	Laburnum anagyroides		●						
C. scoparius			●						
Elaeagnus pungens	Elaeagnus umbellata		●						Sometimes potted-up following bareroot grafting.
Euonymus fortunei	Euonymus europaeus	●	●						
E. alatus		●	●						
E. nanus		●	●						
Fagus sylvatica	Fagus sylvatica		●				●		Often low-worked to avoid swollen union at eye level.
Fraxinus excelsior	Fraxinus excelsior		●						Effectively open-ground budded or grafted.
Hedera helix	×Fatshedera lizei				●	●			Two scions used to fill out head of plant. A specialist Californian product for patios.
Hibiscus syriacus	Hibiscus syriacus		●						Hibiscus syriacus cultivars are sometimes top-worked to produce a novelty tree.

Top-worked tree combinations, *continued*

SCION	ROOTSTOCK	Apical, bare-root, spring	Apical, pot-grown, spring	Hot-pipe	Apical, pot-grown, summer	Side, summer	Side, spring	Budding	COMMENTS
Juniperus horizontalis	*Juniperus virginiana* 'Skyrocket'						●		
J. procumbens							●		
J. sabina							●		
J. scopulorum							●		
J. squamata							●		*Thuja occidentalis* 'Pyramidalis' successfully used for *J. squamata* 'Blue Star'.
Laburnum alpinum	*Laburnum* ×*watereri* 'Vossii'	●							Potted after grafting.
Larix decidua	*Larix decidua*	●	●				●		
L. kaempferi	*L. kaempferi*	●	●				●		
Malus prunifolia	MM106	●	●						MM111 with an interstem of *M.* ×*adstringens* 'Hopa' is used by some growers.
M. sargentii	*M.* 'Bittenfelder'	●	●						
M. 'Royal Beauty'	*M. sylvestris*	●	●						
Picea abies	*Picea abies*						●	●	Produced as miniature standards 45–60 cm tall.
P. pungens							●	●	
Pinus mugo	*Pinus sylvestris*							●	Normally produced as miniature standards 45–60 cm tall.
P. strobus	*P. strobus*							●	
P. sylvestris	*P. sylvestris*							●	
Piptanthus nepalensis (syn. *P. laburnifolius*)	*Laburnum anagyroides*	●							Also open-ground budded or grafted.

SCION	ROOTSTOCK	Apical, bare-root, spring	Apical, pot-grown, spring	Hot-pipe	Apical, pot-grown, summer	Side, summer	Side, spring	Budding	COMMENTS
Prunus ×*cistena*	*Prunus americana*	●	●						Also open-ground budded or grafted. Unsightly union at eye level due to stem overgrowth may develop when top-working many *Prunus* rootstocks, particularly *P. avium* and Mazzard F12/1.
	P. cerasifera	●	●						
	Myrobalan B	●	●						
	P. mahaleb	●	●						
	P. avium	●	●						
P. fruticosa	Mazzard F12/1	●	●						Occasionally top-worked.
P. ×*incam* 'Okamé'		●	●						
P. serrulata	*P. avium*	●	●						*Prunus serrula* used sometimes as an interstem budded onto *P. avium*.
	Mazzard F12/1	●	●						
P. subhirtella	*P. avium*	●	●						Occasionally top-worked.
P. triloba	Myrobalan B	●	●						
	P. cerasifera	●	●						Interstock of *P. nigra* can be used with *P. cerasifera*.
	St. Julien A	●	●						
	Brompton	●	●						
Rhododendron species and cultivars	*Rhododendron* 'Anna Rose Whitney'		●				●		
Ribes nigrum	*Ribes aureum* 'Brecht'		●						Edible currants used to be top-worked in countries like Hungary for commercial production. Make good patio plants.
R. rubrum	*R. aureum* 'Pallagi 2'		●						
R. uva-crispa			●						
Robinia hispida	*Robinia pseudoacacia*		●						Also open-ground budded and grafted.
R. pseudoacacia			●						
Rosa 'Dorothy Perkins'	*Rosa canina* 'Pfänder'		●						Normally open-ground budded.
R. moyesii	*R. multiflora*		●						Occasionally produced as a novelty tree (standard) rose.
R. 'Little Buckaroo'	*R. canina* 'Inermis'		●						

Top-worked tree combinations, *continued*

SCION	ROOTSTOCK	GRAFTING METHOD							COMMENTS
		Apical, bare-root, spring	Apical, pot-grown, spring	Hot-pipe	Apical, pot-grown, summer	Side, summer	Side, spring	Budding	
Salix caprea	*Salix discolor*	●	●						Can also be grafted onto unrooted cuttings of rootstock.
S. hastata	*S.* ×*smithiana*	●	●						
S. helvetica	*S. viminalis* 'Bowles Hybrid'	●	●						*Salix viminalis* 'Bowles Hybrid' is one of the most vigorous and hardy varieties of willow commonly available and more disease resistant than other rootstocks.
S. integra		●	●						
S. purpurea		●	●						Two scions can be put on rootstock to give a fuller head (e.g., side veneer and "rind" graft lower down).
Sophora japonica	*Sophora japonica*		●				●		
Ulmus ×*elegantissima*	*Ulmus glabra*		●						
U. glabra			●						Effectively open-ground budded or grafted.
U. parvifolia			●						

Fruit tree rootstocks

The rootstocks listed below are some of the commonly used rootstocks. The list is not exhaustive. For further information, see the books listed in the bibliography.

SCION	ROOTSTOCK	GRAFTING METHOD				COMMENTS
		Apical, bare-root, spring	Apical, pot-grown, spring	Hot-pipe	Budding	
Actinidia deliciosa (kiwi)	*Actinidia deliciosa* 'Bruno' or 'Hayward'	●			●	Seed-raised rootstocks mainly used.
Citrus	*Citrus jambhiri* (rough lemon)		●	●	●	A vigorous rootstock used for sweet orange, grapefruit, mandarin, and lemon. Good for infertile sandy soils. Tolerant to Citrus Tristeza Virus (CTV), Citrus Exocortis Viroid (CEV), and Citrus Xyloporosis Viroid (CXV). Susceptible to blight and freeze damage. Budding is often by inverted T-budding.
	C. volkameriana (Volkamer lemon)		●	●	●	A vigorous rootstock resistant to mal secco, phytophthora, CTV, CEV and CXV. Slightly more freeze hardy than other lemon rootstocks. Susceptible to blight and the citrus and burrowing nematodes. Not widely used.
	C. macrophylla		●	●	●	Lemon, lime, Valencia orange, and mandarin yield well. Not cold hardy. Susceptible to CTV and CXV.
	C. aurantium (sour orange)		●	●	●	A moderately vigorous, very widely used rootstock worldwide, but sweet orange on this rootstock is susceptible to CTV, reducing its use in areas with this virus.
	Citrus reticulata 'Cleopatra' (mandarin)		●	●	●	Moderately vigorous and drought resistant. Yields are lower than rough lemon but higher than sour orange rootstocks. Tolerant of major virus diseases of citrus and has increased in use in recent years.
	Poncirus trifoliata (trifoliate orange)		●	●	●	Widely used rootstock for Satsuma mandarin and sweet orange.
Diospyros kaki (Japanese persimmon)	*Diospyros virginiana*		●		●	
Diospyros virginiana (common persimmon)	*D. kaki*		●		●	
Eriobotrya japonica (loquat)	*Eriobotrya japonica*		●		●	

Fruit tree rootstocks, *continued*

SCION	ROOTSTOCK	Apical, bare-root, spring	Apical, pot-grown, spring	Hot-pipe	Budding	COMMENTS
Malus domestica (apple)	M9	●			●	Dwarf, very popular rootstock growing to 2.5 m tall. Permanent staking for support required but can be used as an interstock with MM106 to give better anchorage with dwarfing characteristics.
	M25	●			●	Grows very vigorous, over 4 m tall and used for half- and full-standard trees.
	M27	●			●	Dwarf rootstock growing to no more than 2 m tall.
	M26	●			●	Semi-dwarf rootstock growing to 3 m tall, requires support in most sites.
	M116	●			●	Medium, slightly less vigorous than MM106 and good resistance to phytophthora and woolly aphid.
	MM106	●			●	Semi-vigorous growing 3-3.5 m tall.
	MM111	●			●	Vigorous growing 3.5-4.0 m tall.
Persea americana (avocado)	Zentmyer		●	●		Extremely vigorous and highly durable rootstock. Resistant to phytophthora root rot. Not salt resistant.
	Steddom		●	●		Resistant to phytophthora root rot. Salt tolerant.
	Uzi		●	●		Resistant to phytophthora root rot. Produces high, consistent yields. Seedling rootstocks also used widely.
Prunus armeniaca (apricot)	Torinel				●	Semi-dwarf growing to 3 m tall and giving higher yields and larger fruit than other rootstocks. Otherwise seed-raised rootstocks used.
	Myrobalan plum				●	Good for heavy and wet soils.
Prunus avium (cherry)	Colt	●			●	Semi-vigorous growing 4-5 m tall. Compatible with all cultivars and very productive.
	Gisela 5	●			●	Dwarf growing 2.5-3.0 m tall. Very productive.
	Gisela 6	●			●	Similar dwarfing characteristics to Colt. Good productivity.
	F12/1	●			●	Very vigorous growing over 5 m tall. Best form of *P. avium* rootstock.

SCION	ROOTSTOCK	GRAFTING METHOD				COMMENTS
		Apical, bare-root, spring	Apical, pot-grown, spring	Hot-pipe	Budding	
Prunus domestica (plum)	Pixy	●			●	The best plum rootstock for a small garden, growing to the equivalent size of apple M26.
	VVA-1	●			●	A new rootstock from the United States and eastern Europe. Produces a smaller tree than Pixy but still being assessed in the United Kingdom.
	Saint Julien A	●			●	The best general-purpose plum rootstock of comparable vigour to apple MM106. Widely used for apricots, peaches, and nectarines.
	Brompton	●			●	Produces a large standard plum tree.
	Myruni	●			●	A *P. insititia* clone with medium to strong growth. Early cropping with high yields and good fruit size. Does not sucker.
Prunus persica (peach and nectarine)	Seedling peach				●	Research into dwarfing rootstocks is being undertaken and Krymsk seems to have potential.
Pyrus communis (European pear)	Quince C	●			●	Dwarf growing 2.5 m tall. Slightly earlier cropping than other rootstocks.
Pyrus pyrifolia (Asian pear, Nashi)	Quince A	●			●	Semi-dwarf growing 3 m tall.
	Pyrodwarf	●			●	Excellent winter hardiness and resistance to lime-induced chlorosis on calcareous soils. There is no need for support. Good compatibility with all pear cultivars, interstocks are not required. Larger than Quince A.
	Pyrus communis	●			●	Rootstock produced from seed, giving a freestanding tree of full size. The old saying "plant pears for your heirs" comes from this rootstock since it will start fruiting after 8–10 years.
	OH × F87	●			●	Becoming the preferred selection of the Old Home × Farmingdale series that has been developed to give fireblight resistance. Semi-vigorous.
	P. betulifolia	●			●	Rootstock of choice for Asian pears.
Vaccinium corymbosum (highbush blueberry)	*Vaccinium ashei*	●			●	Grafting allows the highbush blueberry to be grown in a wider range of soil types than using cuttings.
	V. arboreum	●			●	*Vaccinium arboreum* produces a more upright, treelike growth habit that has the potential to be used for mechanical harvesting.
Vitis vinifera (grape)		●	●	●	●	There are many rootstocks that have parentage of *V. rupestris, V. riparia,* and *V. berlandieri*. It is important to get local advice for rootstock selection since regional needs and preferences are very diverse. Whip graft onto rooted or unrooted rootstalk cuttings can be used. A one-bud scion is used.

Useful Conversions

Temperature

CELSIUS	FAHRENHEIT
-20°C	-4°F
-12°C	10°F
-7°C	20°F
0°C	32°F
2°C	36°F
3°C	37°F
4°C	39°F
5°C	41°F
10°C	50°F
15°C	59°F
16°C	61°F
18°C	64°F
20°C	68°F
21°C	70°F
22°C	72°F
23°C	73°F
24°C	75°F
25°C	77°F
27°C	81°F
28°C	82°F
29°C	84°F
30°C	86°F
32°C	90°F
35°C	95°F
40°C	104°F
50°C	122°F
60°C	140°F
70°C	158°F
105°C	221°F

Length

MILLIMETRES	INCHES
2 mm	0.08 in.
4 mm	0.16 in.
6 mm	0.24 in.
8 mm	0.3 in.
10 mm	0.4 in.
12 mm	0.5 in.
14 mm	0.6 in.
20 mm	0.8 in.
25 mm	1 in.
30 mm	1.2 in.
40 mm	1.5 in.
50 mm	2 in.
80 mm	3 in.
100 mm	4 in.
140 mm	5.5 in.
150 mm	6 in.
200 mm	8 in.
240 mm	9 in.
300 mm	12 in.
350 mm	14 in.

CENTIMETRES	INCHES
1 cm	0.4 in.
1.5 cm	0.6 in.
2 cm	0.8 in.
2.5 cm	1 in.
3 cm	1.2 in.
5 cm	2 in.
7 cm	2.8 in.
8 cm	3.2 in.
9 cm	3.5 in.
10 cm	4 in.
12 cm	4.8 in.
14 cm	5.5 in.
15 cm	6 in.
20 cm	8 in.
25 cm	10 in.
30 cm	12 in.
40 cm	16 in.
60 cm	24 in.
80 cm	32 in.
100 cm	40 in.
120 cm	47 in.

METRES	FEET
1 m	3.3 ft.
1.1 m	3.6 ft.
1.5 m	4.9 ft.
2 m	6.5 ft.
1400 m	4590 ft.

Weight

GRAMS/KILOS	POUNDS
1 kg	2.2 lbs.
57 kg	126 lbs.
74 kg	163 lbs
90 kg	198 lbs.

Volume

HECTOLITRES	GALLONS
23.4 million hl	618.2 million gal.
84.5 million hl	2232.2 million gal.

Glossary

abscises To separate by abscission, as a leaf from a stem.

adventitious buds Buds that develop from places other than a shoot apical meristem, which occurs at the tip of a stem. They may develop on stems, roots, or leaves.

apical graft Any grafting technique where the top of the rootstock is removed prior to grafting.

approach graft A graft joining two plants so that parts of both, above and below point of union, are temporarily retained.

asexual reproduction Multiplication of plants by vegetative means. Produces genetically identical clonal plants.

axillary bud An embryonic shoot that lies at the junction of the stem and petiole of a plant.

bark All tissues lying outward from the vascular cambium.

budding, bud-grafting Grafting with bud cut from shoot with portion of rind and, in some cases, a small piece of wood. Includes chip-budding (budding in which a small chip of wood and rind with bud replaces similar piece of rootstock) and T- or shield-budding (budding in which a thin slice of rind with wood and one bud is inserted under the rind of the rootstock).

bench graft Graft made while both rootstock and scion are out of the soil, work being commonly done at a bench.

Brix The sugar content of an aqueous solution. One degree Brix is 1 gram of sucrose in 100 grams of solution.

bud-stick A detached shoot from which leaf blades are removed, bearing buds to be used as scions when budding.

bud-wood Shoots used as bud-sticks.

callus Undifferentiated healing tissue arising at wounds.

cambium Layer of actively dividing cells around xylem or wood. New cells arising differentiate to become xylem on inner side and phloem or rind on outer side.

cell tray A compartmentalized tray for growing seed or rooting cuttings. Also called a module or plug tray.

certified stock Stock of plant material (that is, rootstocks or bud-sticks) that has been entered in an official certification scheme, has been inspected, approved, and granted a certificate indicating that it is true to name and of specified degree of freedom from disease.

chimaera A plant or shoot formed of two or more genetically different types of cells. It may arise spontaneously by mutation in bud, or more rarely because of grafting, when it is known as a graft chimaera.

cleft graft Any graft not involving separation of rind from wood of rootstock, more particularly insertion of scion into cleft or split stock.

clone Vegetatively propagated and identical progeny originating from single plant.

coir The fibre obtained from the husk of a coconut used as a substrate for growing plants.

compatibility Ability of rootstock and scion to form good permanent union.

composite scion Scion formed, as in some forms of double grafting, by grafting together two or more scions.

container A pot used for growing plants, often hardy nursery stock. Usually refers to sizes of one litre and above.

cotyledon An embryonic leaf in seed-bearing plants, one or more of which are the first leaves to appear from a germinating seed.

crown graft Top working in which a number of scions are placed around the cut end of limb, giving the appearance of a crown.

cucurbit A plant of the gourd family (Cucurbitaceae), which includes melon, pumpkin, squash, and cucumber.

cultivar A selected plant given a unique name as it has desirable characteristics that are distinct from the species, other cultivars, and are uniform and stable.

cytokinin Any of a class of plant hormones that promote cell division and growth, and delay the aging of leaves.

dedifferentiation Regression of a specialized cell or tissue to a simpler, more embryonic, unspecialized form. Dedifferentiation may occur before the regeneration of appendages in plants.

dormant A state of greatly reduced metabolism in which a plant is alive but not growing.

double-worked Referring to a tree composed of three parts—rootstock, intermediate scion, and upper scion—thus grafted or budded more than once.

dwarfing rootstock A rootstock that induces restricted, but otherwise healthy and productive growth in scions worked upon it.

etiolated Developed without chlorophyll by being deprived of light.

evapotranspiration The process by which water is transferred from the land to the atmosphere via evaporation from the soil and other surfaces and via transpiration from plants.

exudate A substance that oozes out from plant pores.

frame-work To rework an established tree by grafting scions onto existing framework of branches. See *topwork*.

genotype The genetic makeup of an individual organism's inheritable information.

gibberellin Any of a group of plant hormones that stimulate stem elongation, germination, and flowering.

graft (1) To unite two portions of plants by bringing their cambiums into contact and providing conditions for cut surfaces to grow together. (2) Completed operation of grafting. (3) The point of union of rootstock and scion.

graft chimaera A plant arising from the point of union of a scion and rootstock composed of cells from the two plants and showing properties of the two plants.

grafting tape Plastic or adhesive textile tape used to tie and seal grafts in one operation.

grafting wax Graft-sealing compound. Some types are applied hot, others cold, by dipping or with brush, spatula, or fingers.

guide clip A pliable metal clip used to obtain very straight or erect new growth from scion.

guttation The exudation of drops of xylem sap on the tips or edges of leaves of some vascular plants.

hardwood cutting. See *leafless winter cutting*.

head back To cut the rootstock back to "bud" after union has taken place.

heterozygous Having dissimilar pairs of genes for any hereditary characteristic.

high worked A standard or half-standard tree grafted or budded with one or more scions at point of future crotch. Often referred to as top working.

homozygous Having similar pairs of genes for any hereditary characteristic.

hot-pipe A method of applying heat at the point of union when winter bench grafting to promote the graft union to form while retaining the scion buds in a dormant state.

hypocotyl The part of the stem of an embryo plant beneath the stalks of the seed leaves, or cotyledons, and directly above the root.

imbibition The process of water absorption by dry seed before germination.

incompatibility The relation existing between two plants which, when intergrafted, fail to form a complete and lasting union.

inlay graft Graft in which rootstock is headed back and a wedge or sector cut out and replaced by scion cut to fit. See *veneer graft*.

inosculation A natural phenomenon in which the trunks, branches, or roots of two trees grow together. This may occur between compatible species, causing a natural graft, or between incompatible species where both trees will only continue to grow above the union while attached to their own roots.

interstock Scion interposed between a rootstock and another scion. See *double-worked*.

isozymes Each of two or more enzymes with identical function but different structure.

leafless winter cutting Stem cutting of a deciduous woody plant that is taken in winter when it is dormant and leafless. Also called a hardwood cutting.

lignification To turn into wood or become woody through the formation and deposit of lignin in cell walls.

lignin A complex organic polymer deposited in the cell walls of many plants, making them rigid and woody.

line-out To set young plants in rows in open nursery either for further growth, or for budding or grafting.

liner A small container usually to grow on a propagule from the propagation stage until potting into its final container. The term derives from *line-out*.

low-worked Grafted or budded at or near ground level. Sometimes called bottom-worked.

maiden A low-worked tree in which scion has grown for one season. May be referred to as a one-year whip in the United States.

meristematic Undifferentiated tissue from which new cells are formed, as at the tip of a stem or root.

natural grafting A natural phenomenon in which trunks, branches, or roots of two trees grow together. A form of inosculation but the plants are compatible and will continue to grow above the union even if one of them is severed from its roots.

necrotic Death of cells or tissues through injury or disease.

nematode An unsegmented worm with an elongated rounded body pointed at both ends; mostly free-living but sometimes parasitic. May be a vector for virus transmission to plants. Commonly called eelworm.

ovule culture A minute structure in seed plants, containing the embryo sac and surrounded by the nucellus, that develops into a seed after fertilization. Refers to its use in tissue culture.

parenchyma The primary tissue of higher plants, composed of thin-walled cells and forming the greater part of leaves, roots, the pulp of fruit, and the pith of stems.

pathogens A bacterium, virus, or other microorganism that can cause disease.

periderm The corky outer layer of a plant stem formed in secondary thickening or as a response to injury or infection.

perlite A natural volcanic glass similar to obsidian used as insulation or in plant growth media.

peroxidase enzyme An enzyme that catalyzes the transfer of oxygen from hydrogen peroxide to a suitable substrate and thus brings about oxidation of the substrate.

petiole The stalk by which a leaf is attached to a stem.

phelloderm The layer of tissue, often very thin, produced on the inside of the cork cambium in woody plants. It forms a secondary cortex.

phenotype The appearance of an organism resulting from the interaction of the genotype and the environment.

phloem The living tissue that carries organic nutrients in particular, sucrose, a sugar, to all parts of the plant.

photosynthesis The process by which green plants use sunlight to synthesize foods from carbon dioxide and water

polarity Differing physiological reactions as between basal and apical ends of a portion of plant which cause the former to produce roots and the latter shoots, irrespective of orientation in respect to gravity.

polyphenolics Any of various alcohols containing two or more benzene rings that each has at least one hydroxyl group (OH) attached.

propagule Any plant organ or part, as a spore, seed or cutting, used to propagate a new plant.

radicle The first part of a seedling to emerge from the seed during the process of germination.

rework To graft a plant or tree that has already been worked; usually undertaken to change variety of established tree. See *frame-work* and *topwork*.

rind graft Any graft in which the scion is inserted between rind and wood.

rockwool An inorganic fibrous substance that is produced by steam blasting and cooling molten glass or a similar substance and is used as an insulator and as a substrate for plant growth.

root graft A form of bench grafting in which rootstock is a portion of root.

root primordia A root in its initial stage of development, originating from the mature pericycle, located between the epidermis and phloem, of the parent root.

rootstock A plant upon which one or more scions are to be, or have been, grafted.

saddle graft A graft in which a scion is divided at the base like a clothes peg and is fitted to a rootstock, which is cut to form an upward-pointing wedge. There are many modifications.

sap-drawer A branch retained temporarily in a reworked tree to provide nutrition and to maintain activity near or above the level of graft, to give partial shade, or to reduce bleeding.

scion Part of plant used to provide shoot system when grafted upon a rootstock. Sometimes known as cion in the United States.

scion-rooting Development of roots from the scion so that grafted plant has roots arising directly from both rootstock and scion.

scion-wood Shoots from which scions are cut.

side graft Any graft that retains the top of the rootstock until after the graft union has formed.

side veneer graft A graft in which scion is placed on the side of a stem or branch as distinct from end. A section is removed from the rootstock and replaced by a scion with a matching cut.

side wedge graft (or modified side veneer graft) Similar to the side-veneer graft except that no tissue is removed from the rootstock. A thin flap consisting of rind and a small amount of tissue remains attached to the base of the rootstock and this flap is replaced over the scion.

single-worked tree A tree composed of two parts, namely, rootstock and scion.

snag Part of rootstock left temporarily above union and used as support for growth from scion; particularly applicable to budded stock.

solenaceous vegetables Any of several fruits of plants of the family Solanaceae, especially the genera *Solanum* (aubergine, potato, tomato) and *Capsicum* (sweet and hot peppers).

specific fluid A theory of plant growth commonly held from the time of the Greeks to the seventeenth century. It proposed that a given species grows by extracting from the soil a fluid specific to its particular kind (and no other).

splice graft. See *whip graft*.

stembuilder A variety used as an intermediate stempiece between rootstock and scion to provide good trunk for standard trees, or to introduce resistance to disease, or to winter injury. The rootstocks may also serve as stembuilder without any intermediate.

stock plants Selected plants from which seed is to be saved or vegetative material taken for propagation.

sucker A shoot arising below a graft on worked trees.

tannin Any of various compounds, including tannic acid, that occurs naturally in the bark and fruit of various plants, especially *Quercus* (oak) species, nutgalls (galls on oaks that resemble nuts and are high in tannic acid), and *Rhus typhina* (sumac).

tie-in (1) To secure bud or graft with grafting tape. (2) To support growth from bud by tying it to snag or stake.

topwork (1) To rework an established tree by cutting back branches near to crotch and grafting cut branch ends. (2) Used as an alternative term to high working.

totipotency The ability of a single cell to divide and produce all of the differentiated cells in a plant.

turgor The state of turgidity and resulting rigidity of plant cells typically due to the absorption of water.

uncongeniality (1) Mild incompatibility in grafted plants, sometimes shown by large swellings or overgrowths at graft union. (2) Severely dwarfed growth accompanied by symptoms of ill-health or unthriftiness attributable to grafting.

understock See *rootstock*.

union The junction of scion and rootstock.

veneer graft A graft in which part of rind of rootstock without bud is replaced by rind of scion.

whip-and-tongue graft Graft similar to whip graft except that small interlocking tongues are cut in rootstock and scion.

whip graft Graft in which rootstock and scion of similar diameter are joined by matching sloping cut surfaces. Also called splice graft.

work To graft or bud.

worked tree A tree that has been grafted, or budded.

References

Garner, R. J. 2013. *The Grafter's Handbook*. 6th edition, revised and updated by S. Bradley. London: Octopus Publishing Group.

Hartmann, H. T., D. E. Kester, F. T. Davies, and R. L. Geneve. 2013. *Hartmann & Kester's Plant Propagation: Principles and Practice*. 8th edition. London: Pearson.

Lamb, K., J. Kelly, and P. Bowbrick. 1995. *Nursery Stock Manual*. Grower Manual No. 1. Swanley, Kent: Grower Books, Nexus Media.

Lewis, W. J., and D. McE. Alexander. 2008. *Grafting and Budding: A Practical Guide for Fruit and Nut Plants and Ornamentals*. 2nd edition. Collingwood, Australia: Landlinks Press.

MacDonald, B. 2000. *Practical Woody Plant Propagation for Nursery Growers*. Portland, Oregon: Timber Press.

Chapter 1: History of Grafting

Darwin, C. 1883. *The Variation of Animals and Plants Under Domestication*. 2 vols. New York: D. Appleton and Company.

Janick, J. 2010. The origins of fruits, fruit growing, and fruit breeding. In *Plant Breeding Reviews*, vol. 25. Ed. J. Janick. Oxford, England: John Wiley and Sons. 225–321.

Lammerts van Bueren, E. 2007. Values in organic agriculture and their consequences for a process-orientated evaluation of plant breeding techniques. *HortScience* 42: 813.

Lui, Y. 2006. Historical and modern genetics of plant graft hybridization. *Advances in Genetics* 56: 101–129.

Maimonides, M. 1856. *The Guide for the Perplexed*. Trans. M. Friedländer. Reprint of the 3rd rev. ed., 1904. New York: Dover Publications.

Mudge, K., J. Janick, S. Scofield, and E. E. Goldschmidt. 2009. A history of grafting. *Horticultural Reviews*, vol. 35. Ed. J. Janick. Oxford, England: John Wiley and Sons. 437–493.

Reid, J. 1683. *The Scots Gard'ner together with The Gard'ners Kalendar*. Edinburgh: D Lindsay Publisher.

Willes, M. 2011. *Making of the English Gardener: Plants, Books, and Inspiration 1560–1660*. New Haven, Connecticut: Yale University Press.

Chapter 2: Uses of Grafting

Barney, D. L., and K. E. Hummer. 2005. *Currants, Gooseberries, and Jostaberries: A Guide for Growers, Marketers, and Researchers in North America*. Boca Raton, Florida: CRC Press.

Burbidge, F. W. T. 1877. *Cultivated Plants: Their Propagation and Improvement*. Edinburgh, Scotland: Blackwood and Sons.

Calavan, E. C., S. Van Gundy, J. W. Eckert, and E. L. V. Johnson. 1982. Diseases and their control. *California Agriculture* 36 (11): 10–12.

Campbell, C. 2005. *The Botanist and the Vintner: How Wine Was Saved for the World*. Chapel Hill, North Carolina: Algonquin Books.

Erlandson, W. 2001. *My Father "Talked to Trees."* W. Erlandson.

Jackson, J. E. 2003. *The Biology of Apples and Pears*. Cambridge, England: Cambridge University Press.

Oberly, G. H. 1974. *Top-working and Bridge-*

Grafting Fruit Trees. Cornell Cooperative Extension Publication. Information Bulletin 75.

Raulston, J. C. 1995. New concepts in improving ornamental plant adaptability with stress-tolerant rootstocks. In *Combined Proceedings, International Plant Propagators' Society* 45: 566–569.

Chapter 3: Formation of Graft Union

Andrews, P. K., and C. S. Marquez. 1993. Graft incompatibility. *Horticultural Reviews*, vol. 15. Ed. J. Janick. Oxford, England: John Wiley and Sons. 183–232.

Asante, A. K. 1970. Effect of applied pressure on callus formation and its relevance in grafting. *Ghana Journal of Agricultural Science* 32: 145–151.

Atkinson, C. J., M. A. Else, and T. Blanusa, T. 2000. Explaining the rootstock effect (leaflet). Fruit Focus (Annual Conference East Malling).

Barnett, J. R., and A. K. Asante. 2000. The formation of cambium from callus in grafts of woody species. In *Cell and Molecular Biology of Wood Formation*. Eds. R. A. Savage, J. R. Barnett, and R. Napier. Oxford, England: BIOS Scientific Publishers. 155–168.

Bush, R. F. 1995. The pitfalls of grafting. In *Combined Proceedings, International Plant Propagators' Society* 45: 296.

Howard, B. H. 1993. Investigations into inconsistent and low bud-grafting success in *Acer platanoides* 'Crimson King'. *Journal of Horticultural Science and Biotechnology* 68 (3): 455–462.

Howard, B. H., and W. Oakley. 1997. Bud-grafting success in *Acer platanoides* 'Crimson King' related to root growth. *Journal of Horticultural Science and Biotechnology* 72 (5): 697–704.

Howard, B. H., D. S. Skene, and J. S. Coles. 1974. The effects of different grafting methods upon the development of one-year-old nursery apple trees. *Journal of Horticultural Science and Biotechnology* 49: 287–295.

Lamb, J. G. D., and F. Nutty. 1981. *An Introduction to the Grafting of Nursery Stock*. An Foras Taluntais, handbook series 18. Dublin: An Foras Taluntais.

Loach, K. 1981. Propagation under mist and polythene: history, principles, and development. *Askham Bryan Horticulture Technical Course* 21: 23–31.

McPhee, G. R. 1992. Grafting techniques. In *Combined Proceedings, International Plant Propagators' Society* 42: 51–53.

Meacham, G. E. 1995. Bench grafting, when is the best time? In *Combined Proceedings, International Plant Propagators' Society* 45: 301–304.

Moore, R. 1983. Studies of vegetative compatibility-incompatibility in higher plants IV. The development of tensile strength in a compatible and an incompatible graft. *American Journal of Botany* 70 (2): 226–231.

Nelson, S. H. 1968. Incompatibility survey among horticultural plants. In *Combined Proceedings, International Plant Propagators' Society* 18: 343–407.

Santamour, F. S. 1988. Cambial peroxidase enzymes related to graft incompatibility in red oak. *Journal of Environmental Horticulture* 6 (3): 87–93.

Santamour, F. S. 1996. Potential causes of graft incompatibility. In *Combined Proceedings, International Plant Propagators' Society* 46: 339–342.

Santamour, F. S., and P. Demuth. 1981. Variation in cambial peroxidase isozyme in *Quercus* species, provenances, and progenies. In *Tree Improvement and Genetics—Northeastern Forest Tree Improvement Conference Proceedings—1981* 27: 52–62.

Young, P. J. 1992. Fruit tree propagation. In *Combined Proceedings, International Plant Propagators' Society* 42: 61–64.

Chapter 4: Production of Rootstock and Scion Material

SEED

Anonymous. 1979. *The Production and Management of Rose Rootstocks.* Loughgall, Northern Ireland: The Horticulture Centre, Department of Agriculture for Northern Ireland.

Barnett, J. P. 2008. Relating seed treatments to nursery performance: experience with southern pines. In *USDA Forest Service Proceedings* 57: 27–37.

Derkx, M. P. M. 2006. Liquid sorting and film coating: techniques for improving tree seed performance. In *Combined Proceedings, International Plant Propagators' Society* 56: 200–204.

Dictus, N. 2006. Experience using controlling seed moisture content during stratification to improve germination on the nursery. In *Combined Proceedings, International Plant Propagators' Society* 56: 198–199.

Gordon, A. D., and D. C. F. Rowe. 1982. Seed manual for ornamental trees and shrubs. Forestry Commission Bulletin 59. London: Her Majesty's Stationery Office.

Kunneman, B. P. A. M. 1996. Recent research on propagation at the Research Station for Nursery Stock, Boskoop. In *Combined Proceedings, International Plant Propagators' Society* 46: 197–200.

McMillan Browse, P. D. A. 1979. *Hardy Woody Plants from Seed.* Swanley, Kent: Grower Books, Nexus Media.

Schmidt, L. H., and K. A. Thomsen. 2003. Tree seed processing. In *Seed Conservation: Turning Science into Practice.* Eds. R. D. Smith, J. B. Dickie, S. H. Linington, H. W. Pritchard, and R. J. Probert. London: Royal Botanic Gardens, Kew. 281–306.

Wood, T. 1996. The right rootstock for a good graft stick. In *Combined Proceedings, International Plant Propagators' Society* 46: 167–168.

VEGETATIVE PROPAGATION

Jackson, J. E. 2003. *The Biology of Apples and Pears.* Cambridge, England: Cambridge University Press. 88–99.

Leakey, R. R. B. 1985. The capacity for vegetative propagation in trees. In *Attributes of Trees as Crop Plants.* Eds. M. G. R. Cannell and J. E. Jackson. Abbotts Ripton, Huntingdon, England: Institute of Terrestrial Ecology. 110–133.

Lovell, P. H., and J. White. 1986. Anatomical changes during adventitious root formation. In *New Root Formation in Plants and Cuttings.* Ed. M. B. Jackson. Leiden, The Netherlands: Martinus Nijhoff Publishers. 111–140.

Tredici, P. D. 1991. Topophysis in gymnosperms: an architectural approach to an old problem. In *Combined Proceedings, International Plant Propagators' Society* 41: 406–409.

STOOL (MOUND) LAYERING

McMillan Browse, P. D. A. 1980. *Stooling Nursery Stock.* Grower Guide No. 19. Swanley, Kent: Grower Books, Nexus Media.

LEAFLESS (WINTER) HARDWOOD CUTTINGS

Howard, B. H. 1978. Field establishment of apple rootstock hardwood cuttings as influenced by conditions during a prior stage in heated bins. *Journal of Horticultural Science and Biotechnology* 53 (1): 31–38.

Chapter 5: Bench Grafting

Barnett, J. R., and A. K. Asante. 2000. The formation of cambium from callus in grafts of woody species. In *Cell and Molecular Biology of Wood Formation.* Eds. R. A. Savage, J. R. Barnett, and R. Napier. Oxford, England: BIOS Scientific Publishers. 155–168.

Dunn, N. D. 1995. The use of hot-pipe callusing for bench grafting. In *Combined Proceedings,*

International Plant Propagators' Society 45: 139–141.

Hide, D. 2009. Magnolia chip budding. International Plant Propagators' Society, Great Britain & Ireland Region. Winter Newsletter.

Howard, B. H. 1977. Chip budding fruit and ornamental trees. In *Combined Proceedings, International Plant Propagators' Society* 27: 357–368.

Lane, C. 1993. Magnolia propagation. In *Combined Proceedings, International Plant Propagators' Society* 43: 163–166.

Larson, R. A. 2006. Grafting: a review of basics as well as special problems associated with conifer grafting. In *Combined Proceedings, International Plant Propagators' Society* 56: 318–322.

Savella, L. 1977. Poly tent versus open bench grafting. In *Combined Proceedings, International Plant Propagators' Society* 27: 369–371.

Van Allmen, F. 1985. Grafting of *Eucalyptus ficifolia*. In *Combined Proceedings, International Plant Propagators' Society* 35: 137–138.

Chapter 6: Field Grafting

Howard, B. H. 1977. Chip budding fruit and ornamental trees. In *Combined Proceedings, International Plant Propagators' Society* 27: 357–365.

Lane, C. 1995. Propagation and production of *Hamamelis* cultivars in the field by chip budding. In *Combined Proceedings, International Plant Propagators' Society* 45: 149–150.

van Tol, R. W., H. J. Swarts, A. van der Linden, and J. H. Visser. 2007. Repellence of the red bud borer *Resseliella oculiperda* from grafted apple trees by impregnation of rubber budding strips with essential oils. *Pest Management Science* 63 (5): 483–490.

Watanabe, S. 1999. Japanese traditional techniques of plant cultural propagation. In *Combined Proceedings, International Plant Propagators' Society* 49: 124.

Chapter 7: Vegetable Grafting

Burbidge, F. W. 1875. Curiosities of grafting. *The Garden* 8: 460.

King, S. R., A. R. Davis, X. Zhang, and K. Crosby. 2010. Genetics, breeding, and selection of rootstocks for Solanaceae and Cucurbitaceae. *Scientia Horticulturae* 127: 106–111.

Kubota, C., M. A. McClure, N. Kokalis-Burelle, M. G. Bausher, and E. N. Rosskopf. 2008. Vegetable grafting: history, use, and current technology status in North America. *HortScience* 43 (6): 1664–1669.

Lee, J.-M. 1994. Cultivation of grafted vegetables 1. Current status, grafting methods, and benefits. *HortScience* 29 (4): 235–239.

Roberts, I. P. 1891. Herbaceous grafting: a hitherto little practiced and successful method of treating herbs, with curious results. *Scientific American Supplement* 31 (795): 63–69.

Chapter 9: The Future of Grafting

Harford, T. 2013. Hotpants vs. the knockout mouse. *Pop-Up Economics*, BBC Radio 4, 16 January.

Legare, M. 2007. The future of grafting. In *Combined Proceedings, International Plant Propagators' Society* 57: 380–383.

Appendix: Fruit Tree Rootstocks

Davies, F. S., and L. G. Albrigo. 1994. *Citrus*. Crop Production Science in Horticulture 2. Wallingford, England: CABI.

Jackson, D. I., and N. E. Looney, eds. 1999. *Temperate and Subtropical Fruit Production*. 2nd ed. Wallingford, England: CABI.

Rom, R. C., and R. F. Carlson, eds. 1987. *Rootstocks for Fruit Crops*. Hoboken, New Jersey: Wiley InterScience.

Photo Credits

Index

scions, *continued*
  top-work tree combination tables, 196–200
  for whip grafting (winter), 132
scions, dormancy of
  conifers, 123
  for graft success, 54, 59
  hot callus pipe maintaining, 63, 87, 109, 110–111
  scion collection and, 90–92
  storage practices, 88, 91, 92, 105
  timing the graft, 58–60, 96–98, 132
sclerenchyma cells, 83
screening for disease indexing, 48–49
sealing, 63
secateurs, 79, 105, 131
sectorial chimera, 31
seed dormancy, 68–69, 71–73
seedling peach rootstock, 203
seedling specifications, 66
seed orchards, 35, 50
seed-raised forestry trees, 35
seed-raised rootstock. *See* rootstock from seed
seed viability, 73–77
Selenicereus, 168–169
Sequoia 'Aptos Blue', 193
Sequoiadendron giganteum, 193
Sequoia sempervirens, 193
shade, 63–64, 98, 99
shapes, grafting into, 42
Sharrock, Robert, 26–27, 28
shield graft, 140–141
shot buds, 145
shrubs
  cold callusing for, 99, 108
  grafting overview, 34, 42, 43
  hot-pipe callus, 113

micropropation, 49–50
scion production, 90
summer grafting, 115
Sicyos angulatus, 158
side graft, 95, 98, 109, 115, 117, 179
side veneer graft
  for conifers, 123, 124, 126–128
  and hot callusing methods, 109, 110
  overview, 95, 138–139
side wedge (modified side veneer) graft, 95, 116, 126–128, 139
Siebold, Philipp Franz von, 146
sieve tubes, 54, 56
silicone clips, 163
slipping, 59
soil diseases, 78, 156, 157, 161–162
soil pH adaptability, 41
Solanum habrochaites, 159
Solanum integrifolium, 159
Solanum lycopersicum, 156, 159
Solanum lycopersicum × *S. habrochaites*, 161–162
Solanum melongena, 156, 159
Solanum nigrum, 31
Solanum pimpinellifolium, 159
Solanum proteus, 31
Solanum sisymbriifolium, 162
Solanum torvum, 159, 162
Solanum tuberosum, 157
Solanum tubingense, 31
solenaceous vegetables, 157
Sophora japonica, 200
Sorbus, 61, 98, 104, 147, 154
  'John Mitchell', 193
  'Wilfrid Fox', 193
Sorbus alnifolia, 193

Sorbus aria, 193
  'Lutescens', 38
Sorbus ×*arnoldiana*, 193
Sorbus aucuparia, 68, 70, 74, 98, 105, 107, 193, 196, 197
Sorbus commixta, 193
Sorbus domestica, 193
Sorbus hupehensis, 193
Sorbus ×*hybrida*, 193
Sorbus intermedia, 38, 181, 193
Sorbus megalocarpa, 193
Sorbus meliosmifolia, 193
SO4 rootstock, 36
specific fluid principle, 26–28
splice graft. *See* whip graft
spring grafting, early, 97–98. 179. *See also* cold callusing (early spring) method; hot callusing (early spring) method
spruce, 125
squash, 156, 158
standards for nurserystock, 66
star-cucumber, 158
Steddom rootstock, 202
stem diameter, 60, 66–67
stenting, 84
stock plant cuttings, 85–88
stool bed layering, 77, 79–81
stratification, 68, 69, 70–74
strawberries, 50
suckering, 55, 175
sulphuric acid scarification, 68, 142
summer grafting, late, 96, 97, 115–117, 179
supplies and tools, 131–132
sycamore, 13, 14, 48, 49
Syringa, 99, 108
Syringa josikaea, 194
Syringa oblata var. *dilatata*, 194

About the Author

ED ROBERTSON

PETER T. MACDONALD teaches plant propagation and nursery stock production at Scotland's Rural College (SRUC) in Ayr and Edinburgh. He is a former director of the International Plant Propagators' Society and has researched propagation techniques for industrial applications. He teaches and runs workshops for the National Trust for Scotland's School of Heritage Gardening at Threave Estate.